# The End of Eden

# The End of Eden

*Wild Nature in the Age of Climate Breakdown*

## Adam Welz

BLOOMSBURY PUBLISHING

NEW YORK · LONDON · OXFORD · NEW DELHI · SYDNEY

BLOOMSBURY PUBLISHING
Bloomsbury Publishing Inc.
1385 Broadway, New York, NY 10018, USA

BLOOMSBURY, BLOOMSBURY PUBLISHING, and the Diana logo are trademarks
of Bloomsbury Publishing Plc

First published in the United States 2023

Bloomsbury Publishing Plc does not have any control over, or responsibility for,
any third-party websites referred to or in this book. All internet addresses given in
this book were correct at the time of going to press. The author and publisher regret
any inconvenience caused if addresses have changed or sites have ceased to exist,
but can accept no responsibility for any such changes.

ISBN: HB: 978-1-63557-522-4; EBOOK: 978-1-63557-523-1

LIBRARY OF CONGRESS CATALOGING-IN-PUBLICATION DATA IS AVAILABLE

2 4 6 8 10 9 7 5 3 1

Typeset by Westchester Publishing Services
Printed and bound in the U.S.A.

To find out more about our authors and books visit www.bloomsbury.com and
sign up for our newsletters.

Bloomsbury books may be purchased for business or promotional use.
For information on bulk purchases please contact Macmillan Corporate and
Premium Sales Department at specialmarkets@macmillan.com.

*To my friends and family members—especially Sarah—
whose enduring support made it possible for me to write this book.*

*And in memory of Stefan Neser and David Burg, whose deep
knowledge and contagious love of wild creatures helped steer me
into all sorts of interesting and productive trouble.*

*Their menschiness lives on.*

# CONTENTS

# INTRODUCTION

I T'S A BEAUTIFUL day in New York City in the early 2020s. Spring is breaking out across the metropolis and you're taking a stroll in a large, green park. Many of the people in the park are attached to a portable computer of some kind—that's quite a new thing—but otherwise the scene around you appears much the same as it would have on a random spring morning a couple of generations ago: lawns, trees, shrubs, paths filled with joggers and dog walkers, noisy kids in the playground. Squirrels. There's a bit less litter and less fear of crime than thirty or forty years back, but to most people the park is essentially unchanged.

Except it's not. Look closely, and you'll discover that many plants are carrying clumps of shriveled brown mush in place of the flowers that they would normally be showing off at this time of year. They were tricked into flowering weeks earlier by a spell of unusually warm, springlike days that swept across the region when the calendar said it was still winter. Freezing weather returned soon after, though, and destroyed their blooms. Because their flowers appeared long before the emergence of the spring insects that normally pollinate them, the plants have failed to produce seed and the insects now face starvation.

Fungal diseases like Southern Blight, spurred on by higher-than-normal temperatures, are also spreading rapidly among perennial shrubs, and the park's Azalea Lace Bugs—insects that destroy the leaves of azaleas and their relatives—have started breeding early. They'll produce four or five generations this summer instead of the usual three, and unless park workers double down on pest control, there will be a lot of wrecked azaleas by fall. Look closely at the stems of the park's lilacs, and you'll see groups of whitish bumps, populations of a strange insect called White Peach Scale. It's a pest that used to be confined to the southern United States, but it's recently moved into New York.

Ragtag groups of binocular-wielding birders, who have been scouring the park for spring rarities since dawn, are seeing slightly fewer migrant warblers than last year, when they saw slightly fewer than the year before; in fact, the number of migrating songbirds passing through here has dropped by about a third since 1970.* Insect numbers have dropped precipitously, too.

But a handful of new bird species have appeared in recent years, like the Black Vulture, a large scavenger with a black, wrinkly face and sharp, hooked bill. It used to be confined to the warm southern states but is now often seen soaring over New York. The Anhinga, or Devil Bird, is a large fish-eating species that's common in hot, wet parts of the Americas—think Florida and the Amazon Basin—but in April 2023 one flew into the city and stayed for weeks. A small number of species that formerly migrated south in the winter, like the American Robin, are now staying in the park year-round. They're breeding ever earlier in the spring and their numbers are rising.

Many of the park's largest and strongest-looking inhabitants, its big old trees, some of which have grown here for centuries,

---

*   The total number of birds in North America dropped from about ten billion in 1970 to about seven billion in 2018.

have been damaged or obliterated in recent years. They've been progressively weakened by a series of droughts, which have been longer and drier than before. Trees have been broken and toppled by more frequent, increasingly intense thunderstorms and blizzards as well as city-blitzing hurricanes like Irene and Sandy. They're being chewed through by wave upon wave of new insect enemies like the Emerald Ash Borer, the Asian Long-horned Beetle, the Southern Pine Beetle, and, most recently, the Spotted Lanternfly, which are moving up from southern latitudes thanks to warming temperatures in New York. The presence here of at least one of the new pests can be linked to extreme weather: in 1985, Hurricane Gloria swept some small, sap-sucking insects called Hemlock Woolly Adelgids from the southern United States all the way to the Northeast, where their species established itself and has been attacking Eastern Hemlock trees ever since.)

As the park's trees die, their carcasses are trucked away for burning or chipped into tiny fragments to make footpaths. The spaces they leave in the sky, the soil, and the ecosystem rapidly fade from human memory. The park's gardeners don't always know what to replace them with because the weather has become so weird and unpredictable in recent years that it's hard to tell which species will survive. But gardeners have noticed that plants from warmer climes like crape myrtles, Cherrybark Oak, and Japanese flowering camellia—species that would not have survived the local winters just a couple of decades ago—seem to be doing quite well.

This scene is not fantasy. It's a synthesis of actual changes happening in parks and natural areas in and around New York City as I write. One thing that's common to all these changes—besides the fact that many park visitors haven't paid much attention to them—is the dramatic recent increase of energy-trapping gases like carbon dioxide and methane in the atmosphere, which is forcing up the Earth's average air, land, and ocean temperatures. It's sometimes difficult to comprehend how this threatens wild species throughout the world and how thoroughly it is disrupting, unraveling, and impoverishing ecosystems not just in the Arctic, with its now-iconic melting ice sheets and starving Polar Bears, but everywhere.

If Earth was the size of an apple, the atmosphere would be about as thick as the apple's skin. Since the beginning of the Industrial Revolution, in about 1760, humans have been performing an uncontrolled experiment on this fragile, life-giving cocoon. We've decided to burn incomprehensibly many gigatons of fossil fuels and spew their combusted remains into the air while simultaneously destroying millions of acres of forests, savannas, and prairies for industrial agriculture and ever-sprawling roads and cities, thus releasing even greater volumes of energy-trapping gases into the atmosphere.

This is wreaking havoc on nature from the largest to the smallest scales. Extreme weather events are intensifying, megadroughts are afflicting vast regions for years, and megafires are burning up millions of acres with breathtaking speed. But species and ecosystems are also being eroded and rearranged more subtly as local microclimates shift and change, forcing smaller, less-noticed life forms to evolve, move away, or wither into extinction; these intimate ecological breakups and breakdowns are of no less consequence than the so-called natural disasters that generate dramatic headlines, and they're happening all around us.

We're creating an all-penetrating phenomenon that is scrambling ecosystems and triggering cascades of chaos throughout the biosphere. It's something like a monster from ancient Greek mythology that, as it embraces the world, sprouts thickets of writhing necks from which horrifying, sharp-toothed heads emerge. Looking into its faces can turn humans to stone.

This is not the first time a miasma of this kind has enveloped our planet. Energy-trapping gas concentrations and global temperatures have spiked before during the four billion years of life on Earth: natural events like colossal, long-lasting volcanic eruptions and massive meteorite impacts have on a few rare occasions triggered huge climate shifts and devastating mass extinctions that are laid down hard in the fossil record.

But the current rise is different: not only is it caused by humans but data indicate that gas concentrations are going up far faster than in any previous epoch. We've pushed atmospheric carbon dioxide from its natural level of about 280 parts per million to over 420 parts per million in just ten human generations, and we're increasing it further still. There is now more $CO_2$ in the air than there has been for at least three million years.

Since our own species evolved into its current form about two hundred thousand years ago, this means that we're creating an atmosphere profoundly unlike any we—and huge numbers of other species—have lived in before. We are making and breathing changed air.

It's also noteworthy that life achieved its peak of diversity about two hundred thousand years ago. Never did more different species live together on Earth than at the dawn of the human era. Our species emerged in an Eden—not the sanitized, fantastical Eden of religious books, but the most wondrous fabric of living things that our planet has ever been clothed in.

"Climate change" is not my favorite term for the phenomenon at the center of this book, although I sometimes use it. For one thing, it's contradictory: "climate," as we learn about it in school, is constant and predictable, the antithesis of change. Placing these opposites together makes the term subconsciously feel meaningless or perfidious to many people. "Change" also implies reversability: that if the climate is changing, it can easily return to its previous state. This is misleading because many of the shifts and transformations that we have already observed are effectively irreversible and will become increasingly entrenched the longer harmful emissions continue.

I think a term like "climate breakdown" better communicates the current situation. "Breakdown" acknowledges that climate is a type of definable pattern or structure that is breaking apart and not easily reassembled. Its cause—human activity—is well understood, and the word is serious and directional; it doesn't refer to a random, temporary variation in the weather. "Global weirding" is another good one, as it conveys the novelty and strangeness of the climate crisis.

Climate breakdown seldom operates alone: it isn't the only force driving ecosystem collapses and species extinctions. Most of the damage we're seeing is caused by it acting synergistically with other harmful outcomes of human activity, including pollution, habitat destruction, and invasive species. Climate breakdown can inflict further harm on ecosystems already damaged by other forces; it can amplify other forces' impacts on the natural world.

For example, the destructive changes in New York parks described above are occurring at the same time as widespread insect population collapses are being recorded around the globe. The causes of these are not yet well understood—they may primarily be driven by pesticides and habitat destruction—but climate breakdown is almost certainly a contributing factor. In the winter of 2006–2007, researchers found a Eurasian fungus in New York that causes a virulent disease in bats called White-nose Syndrome. It was almost certainly carried there inadvertently by humans, and since then it's spread across North America and killed an enormous number of bats. (Shockingly few bats flit over New York today.) Scientists don't think that its spread has been significantly influenced by climate breakdown, but it might be more lethal in certain areas because of changes in local climates.

———

Science meanders toward greater truth in fits and starts, sometimes losing its way for a while. Many of its revelations are provisional, waiting to be superseded by newer, deeper, more rigorous research. It's nonetheless a powerful and vitally important tool for understanding the era we're entering. Although no one has a complete idea of how climate breakdown will transform the living communities of the biosphere in ten, fifty, a hundred, or a thousand years from now—because it's impossible to make detailed predictions about such complex, interactive systems—scientists and other careful observers are starting to understand some of the new ripples and breaks that are running through wild species and ecosystems. It's challenging work, but by striving to understand the reasons for the changes we're already seeing, they're beginning to be able to make useful predictions about the likely future "nature of nature" on this heating planet.

This book is the result of my efforts to look into the terrifying faces of the new climate behemoth without turning myself to stone. I've chosen to tell the stories of a few species in a few places to illuminate some of the important ways that our gas emissions are fracturing and reshaping the entire natural world. Readers should assume that if a climate-related force is affecting one species in a particular place, it's probably having similar effects on other species elsewhere.

The facts, scenes, and expositions within these pages flow from personal experience, experts' observations, and scientific research, although readers require no specialist knowledge; this book is

written for everyone. I hope it brings you closer to the wild creatures of our tarnished Eden as they try to survive this extraordinary, uncertain period in our shared history and inspires you
to act with intelligence and courage for their persistence—their
continued thriving—with us.

# Energy, Water & Time

T he sun pulses up over the distant crests of the Sierra Nevada
range into an open sky, its direct light drawing out the Mojave
Desert's shapes in subtle pink-yellows against nooks and splashes
of just-born shadows. There's no wind this fall morning, so the
plants are as noiseless as the weather-worn hills made of pale, old
rock inexorably degrading into dry, gravelly sand that fills the flats
and washes between them. Desiccated grasses and the papery
corpses of annual herbs and flowers add an unconvincingly soft
layer to the surface—they germinated, rooted, grew, and died
some time ago, but they've not rotted away.

Every few paces (sometimes more than a few) a larger, more-
robust, apparently more-alive plant holds itself up above the dead
layer. There are just a few different types of these, and although

some are covered with deep-green foliage, none have soft, broad, friendly leaves. Most look quite tough, such as the California Juniper with its rough, shredded bark and tiny, tightly packed, scale-like leaves, and the Mojave Yucca, a small unbranched tree topped with a dense crown of long, dagger-like matte fronds. Some are not just tough but absurdly hostile, like the Silver Cholla cactus, which is completely covered in a dense carpet of silvery-barbed two-inch thorns that bunch hot into your skin if you're unwary enough to brush against it. (Wise desert people carry pliers with them, to rip out cholla thorns. Painful but effective.)

The tallest plant here is the Joshua Tree, which grows in open groves of dozens to thousands of individuals. Joshua Trees can be up to forty feet tall and centuries old, but age grants them no dignity. They don't seem to know which way is up, their shaggy branches extending in whichever direction like the limbs of exuberant, many-armed dancers frozen in mid-leap.

The first impression of a Mojave Desert morning like this one is of bright, eerie stasis. There are living plants here, yes, but they're unmoving. They obviously grew, but they don't seem to be growing, at least not for now.

Give this strange place a few more minutes, though. Shifts and flickers of motion will begin to reveal the animals here: a small Desert Cottontail rabbit hops out from under a Creosote Bush and stops, seemingly lost in thought, its pale-brown fur soft and immaculate, its large eyes dark, wet, and round. Farther off a much larger jackrabbit lopes casually down a wash, long ears up for danger. It disappears into the scrub as a slim Coyote trots into view, quick, efficient, nose to the ground, sniffing and pausing now and then, bright eyes checking on you constantly. Some sort of mouse skitters almost invisibly through the dry vegetation at your feet.

And birds. So many. A squad of Gambel's Quail runs helter-skelter through the scrub, alert but directionless, like lost little avian clowns who want to move together but can't decide where to go. Cactus Wrens chatter from their territories—they're bold, smart, and so much larger than the other wrens of North America. A tiny gray Black-tailed Gnatcatcher searches deep in a shrub for insects as a young Red-tailed Hawk launches confidently from a high rock; she's seen a small rodent in the open, and her crop is empty.

Black-throated Sparrows, neat small creatures with sharp black bibs, are among the most common birds here. They move in groups, low to the ground, pecking up seeds and the occasional insect if they find one. They're constantly in contact with low, sharp tweets, keeping their flocks together, always alert for danger. If they come across a snake or a small predatory bird, they gather round and chatter incessantly, warning all the other birds around of its presence. They're most active early in the morning—their search for food begins before sunrise—and once the day's heat really begins to burn during the mid-morning, they flutter toward the densest, shadiest shrubs they can find, where they try to do very little. Almost every animal in the desert seeks out shade and quiets down as the sun rises to its zenith.

As the afternoon lengthens into evening and the air begins to cool, animal activity picks up again. A keen-eyed Northern Shrike surveys the ground from the top of a thorny shrub, a miniraptor searching for small prey as White-tailed Antelope Squirrels creep between patches of cover in search of one last morsel for the day. Lizards begin to reluctantly pull themselves into safe refuges for the night. As bluish shadows advance and the shrinking areas of still-sunlit land turn flat pale pink, a male Desert Bighorn Sheep, crownlike horns curled around his head, emerges as a proud silhouette on the top of a boulder-strewn hill. He follows a ridge down, hooves gently and surely thudding on the rock, until he reaches a broad, sandy wash, which he calmly crosses to begin his climb up the next boulder-strewn hill.

This place is alive, but now, in the early 2020s, much less alive than it was a century ago. There are far fewer species of birds; the Mojave's avian community is in collapse, but for no immediately obvious reason. Most of the Mojave's natural habitat seems healthy. It's barely marked by humans, with little evidence of toxic pollution, very few ecosystem-disrupting invasive species, and few roads, houses, mines, or factories.

The vanished birds have left no signs of their recent presence behind. We only know about them (and can be sure of the current collapse) because of the foresight and efforts of a pioneering naturalist-ecologist and his students and supporters, who, more than a century ago, foresaw that there would be huge changes in California's fauna and flora and set about to create a baseline data set so we, today, could know what it was like then.

Joseph Grinnell was born in 1877 in Oklahoma, the eldest child of a medical doctor who worked for Native American agencies. The family moved extensively around the United States in his youth. Joseph showed a keen interest in nature and began collecting and preparing specimens of wild creatures in his adolescence. By his mid-teens the Grinnells had settled in Pasadena, California, where Joseph deepened his knowledge of wildlife and nurtured an adventurous spirit; in his late teens and early twenties he made two long trips to Alaska, where he observed bird behavior, collected specimens, and wrote scientific articles.

Grinnell received his master's degree from Stanford University in 1901 and completed his doctorate in zoology at the same institution in 1903. It was during this time that he conceived of a detailed list and bibliography of the birds of California, a project that he worked on intermittently for the rest of his life. His particular interest was the distribution of birds and mammals in this large western state: Where was each species found, and why? In 1904 he and his students began carefully surveying different habitats in the state for their birds and small mammals. He supported himself by teaching at the Throop Polytechnic Institute (today the California Institute of Technology).

In 1907 Grinnell had a chance meeting at Throop that would change his life and indeed the science of ecology. A fabulously wealthy heiress, Mary Montague Alexander, arrived at the biology department. Her parents belonged to two of Hawaii's richest families, sugar dynasties with huge landholdings. She had no need to work and a keen interest in travel and zoology.

Alexander was at Throop to seek expert advice on a trip to Alaska she was planning. She spoke with Grinnell, who invited her home to show her his specimen collection. The heiress and the scientist quickly developed a strong friendship founded on their mutual interests—he visited her home to see her specimens, too—and they soon made plans for a great natural history museum in California, a West Coast rival to the Smithsonian.

In 1908, while Alexander was planning a second Alaska expedition, she asked Grinnell to recommend participants. He sent her a list of names, all men. She replied sharply: "Am rather relieved you could not recommend a lady for our trip, though regret your evident contempt of women as naturalists." She asked around more widely and found Louise Kellogg, an unmarried thirty-year-old member of a moderately wealthy Californian merchant family, to join her. They became close in Alaska and lived the rest of their lives together in a happy "Boston marriage," Kellogg joining Alexander on many of her subsequent expeditions.

On her return from Alaska, Alexander put up the money to establish the Museum of Vertebrate Zoology on the University of California campus in Berkeley. She insisted that Grinnell be the museum's director, and they jointly (and presciently) agreed that the new institution should focus on documenting California's flora and fauna before much of it was destroyed by the rising number of human immigrants to the state, who were busily turning huge tracts of it into farms and cities. Grinnell wrote that the value of the museum's work would not be realized "until the lapse of many years, possibly a century, . . . [when] the student of the future will have access to the original record of faunal conditions in California and the west, wherever we now work."

Between 1904 and 1940 Grinnell and his students—sometimes joined by Alexander and Kellogg themselves—recorded and collected amphibians, reptiles, birds, and mammals from over seven hundred locations across California and nearby states with the aim of documenting the vertebrate fauna of all habitats. Grinnell developed a rigorous system of note-taking; each collected specimen and survey site was carefully and systematically documented. His team preserved more than a hundred thousand animal specimens, generated seventy-four thousand pages of field notes, and made more than ten thousand photographs of animals and their habitats.

During the survey Grinnell became ever more interested in how species evolve in the conditions of particular places, and how one species can contain many slightly different subspecies or local types, each developing unique adaptations to local environments and climates. This thinking was controversial at the time; although most zoologists accepted Darwin's idea that each species had not been created as is by God but had evolved into being, they had not yet accepted that species were continuing to evolve and diversify, or that evolution could be observed as it happened.

Grinnell was also one of the earliest users of the term *ecological niche* to define a species' place or role in a habitat or ecosystem. Grinnell observed that each species had particular physical and behavioral attributes that allowed it to live in particular environments and fulfill particular roles in those environments. A bird species may have a bill and digestive system well-suited to feed on seeds, and thick feathering to deal with extreme cold; it would fill the niche of high-altitude seed-eater. The Grinnellian niche, as it has become known, is largely determined by the climate and physical environment of a species, the set of physical parameters within which a species can live and successfully reproduce over time.

Later, other ecologists observed that niches are often constrained by competing species; a species might theoretically be able to live within the entirety of a particular region with a particular climate, but in fact it only occupies a small part of that region because other species outcompete it in some places. Today many ecologists speak of a species' *fundamental niche* and its *realized niche*. The fundamental niche defines the limits of the area and role that a species can theoretically inhabit in the absences of competitors. The realized niche is the area and role that it actually inhabits in an ecosystem when other, competing species are present.

Grinnell led the Museum of Vertebrate Zoology and the surveys until his death in 1939. His work was carefully preserved, and the museum continued under new leadership with more of Alexander's money (she died in 1950, at the age of eighty-two).

In 2014 a group of scientists at the University of California, Berkeley, began to resurvey many of Grinnell's sites, following his protocols as closely as possible, to see what had changed in the intervening decades. Their results from the Mojave Desert, which they began to publish in 2019, are stunning. On average, survey sites had lost 43 percent of their breeding bird species between Grinnell's original surveys and the recent resurveys. Put starkly, the Mojave has lost nearly half of its breeding bird species in just a few decades.

The new research also showed that species disappearance wasn't random. Certain types of birds, inhabiting particular niches, have done better than others, which gives valuable clues as to why they're gone. Larger, carnivorous birds, like falcons, have fared especially poorly—most large predatory species that bred here a century ago are now much reduced in number and no longer reproducing. Smaller, desert-adapted species generally have not declined as much.

The usual reasons for species extirpations don't apply to most of the Mojave. Habitat destruction has been minimal. Birds haven't been overhunted. Invasive species haven't radically changed ecosystems here. We can't say exactly how each of the "lost" Mojave birds blinked out from its habitat because no one carefully tracked them as they faded away over the last century. But halfway around the world, in southern Africa, we can watch the process of dryland bird decline and extirpation play out in detail, right now.

———

Much of the central region of southern Africa is occupied by the Kalahari, often called a desert but in truth more of a dry savanna. Its soft red sand often forms large dune fields, almost bare in dry seasons but clothed in dense, soft sheets of green grasses interspersed with wildflowers after rain. A few species of deep-rooted trees are tough enough to grow along its few, mostly dry river courses, the best known being the craggy, umbrella-shaped Camelthorn, a tough-barked, hard-wooded cliché of an African thorn tree.

Old Camelthorns are favored natural nesting places for Southern Yellow-billed Hornbills, birds with black-and-white bodies the size of small chickens, strong legs, foot-long tails, and dramatic, thick, four-inch-long bright-yellow bills that curve slightly downward. Yellow-bills are lively and inquisitive, using their large pale-yellow eyes, shaded with long, dark eyelashes, to inspect anything new in their environments. They're monogamous and form extremely strong pair bonds, a requirement for their unusual breeding system, which involves long periods of incarceration.

In spring, male Yellow-bills begin to call around their territories, which are centered on cavities in the trunks of large Camelthorn trees—their nest sites. Soon each male is approached by an eligible female. He responds by bringing her tasty gifts of large insects, lizards, and small rodents—Yellow-bills' normal diet—and they erupt together in loud, rising, bubbling song duets, a *Kwok-kwok! Kwok-kwok-kwok!* that can be heard for a mile or more. After two weeks of courtship the female enters the nest hollow, and together they begin to close her in, each partner collecting their feces to construct a mudlike wall over the cavity's opening.

The feces wall dries and hardens fast, forming a strong barrier to protect the nest from predators. The birds leave a slit in the wall just wide enough for the female to receive food through. As soon as she is locked away in the near darkness, she begins to shed her feathers, a molt that leaves her temporarily flightless. She also lays her eggs— four is normal—and begins her month-long incubation of them.

Once his mate is closed within the nest, the male must work extra hard to keep them both alive. He must collect food for himself and for her, which means flying much more than usual. He must constantly keep a wary eye out for the Kalahari's many predators as he searches for insects, small mammals, and tasty reptiles.

But more fundamentally, he must carefully regulate the amount of energy and water in his body. Too much or too little of either, and he dies.

He must work with and against the environment to accomplish this. In the early mornings, evenings, and at night the air is cooler than his body, so he must prevent energy from leaving his body; it's as if the environment is trying to steal it from him. During the hot parts of the day the air is warmer than his body, so he must offload energy to keep himself relatively cool; during this time, it's as if the environment is trying to force too much energy into him.

Managing energy—which is often tightly linked to managing water, because organisms can use water to cool down—is a full-time job for every living thing. We often think of life being sustained by matter, like food, and death being caused by disease or violence wrought by external agents. But life is also fundamentally sustained by formless, invisible energy, so you can say that every organism stays alive by engaging in an endless negotiation, a continuous dance of give-and-take, with the environment over energy. Sometimes this negotiation turns into war, and the front line of this war is the organism's body wall.

———

Conventional physics holds that the universe is formed from two fundamental things: matter—the substance from which all material is made, aka stuff—and energy, commonly defined as the capacity to do work.

Energy is strange. Even though it has been intensively studied, physicists still can't really say what it *is*, in part because it can't be measured directly. We can, though, measure what energy *does*, and so have figured out useful things about it, including that it cannot be created or destroyed, but—as per the first law of thermodynamics—only changed from one form to another.

Physicists classify forms of energy into two high-level categories: potential energy, the energy a body has because of its position, and kinetic energy, the energy a body has because of its motion. Examples of potential energy are gravitational energy and the energy held in chemical bonds between atoms. Kinetic energy includes the energy of motion itself as well as radiation energy, which is energy that moves at the speed of light through space or matter in forms that can be described as particles or waves. (A ball on a shelf has potential energy, which is transformed into kinetic energy when you knock it off the shelf and it falls toward the floor.)

There are other forms of energy as well, but the one that is perhaps most important to understand for the purposes of this book is thermal energy, the form that many people call heat.[†]

All atoms and molecules are constantly in ultramicroscopic motion—vibrating, rotating, and bumping into each other. Thermal energy is the energy that an object or system has by virtue of this movement of the atoms and/or molecules it's made of. Thermal energy is also what gives things their temperature: the more an object or system has, the more its atoms and/or molecules move

---

[†] Many writers assume that thermal energy is the same thing as heat, but heat can in fact mean different things to different people. In the field of thermodynamics (the study of the relationships between different forms of energy), heat refers to the transfer of energy between two systems or entities. Crudely speaking, as far as thermodynamicists are concerned, heat is energy in motion between things, not energy in things. But the word *heat* is also used to mean energy in general by the International Union of Physiological Sciences, which represents biologists who study the functioning of living organisms and their parts; to them, heat is energy, not a particular form of energy. And to many nonscientists, something contains heat if it feels hot, which can be misleading. Picture two identical metal cups: one cup is naked, but the second is wrapped in insulating fabric. If I fill both cups with hot water of the same temperature, each cup will (initially, at least) contain the same amount of thermal energy, but the naked cup's outside will feel hotter to the touch than the insulated cup. To many people, the naked cup will seem to contain more so-called heat, even though it in fact contains less thermal energy because, being uninsulated, it is losing thermal energy ("cooling down") faster than the insulated cup. When you touch a cup, you are not actually feeling the thermal energy inside the cup, you are feeling the thermal energy that is moving *out of* the cup; but it's easy to assume that the so-called heat you feel on the outside of an object is proportional to the so-called heat inside it. To complicate things even more, thermal energy is also sometimes called "heat energy" or "sensible heat" by different scientists. To cut down on confusion, I'll usually use the term *thermal energy* in this book instead of *heat* or other similar terms.

around, and the higher its temperature. The less thermal energy an object or system contains, the lower its temperature.‡

Temperature affects every physical thing. It influences how energy moves in the world, how it moves within the bodies of living things, and how it moves between the world and the bodies of living things. It determines what living things can do, and whether they thrive or perish. It is one of the fundamental controlling factors of life.

The internal temperatures of all living things need to stay within particular ranges, which is to say that if they contain too little or too much thermal energy, they die. Some organisms need to maintain their body temperatures within very narrow limits, and others can survive huge variations. Also, organisms can have different temperature tolerances at different stages of life; for example, mature plants of many species will die if their internal temperature drops toward freezing, while the seeds of the same species may survive far below freezing temperatures for years.

Organisms are constantly exchanging energy with their environment. They do this in three different ways: by gaining or losing kinetic energy (in the forms of light or thermal energy), by taking in or discharging chemical potential energy held in molecular bonds, and via work. All of these change the internal energy of an organism, and all these means of energy exchange are connected because energy can be transformed into different types.

One way an organism can gain energy is from light, which is a form of radiant kinetic energy. For example, a lizard can sit in the sunshine and its scales and skin can transform the radiant energy

‡ There are in fact various definitions of temperature, some of which require a solid understanding of theoretical physics to grasp, but for our purposes it's sufficient to understand temperature as a measure of the thermal energy something contains.

(light) that falls on its body into thermal energy ("heat"), which then passes into the lizard's body and raises its internal temperature. Or an animal can take in chemical potential energy by eating food. When it digests the food, it metabolizes the nutrient molecules, transforming chemical potential energy into kinetic energy; thus warm-blooded animals like us can use the energy in food to warm their bodies. Organisms can also exchange energy with the world via work; when a kangaroo extends its legs to push against the ground and jump, it is transferring—losing—energy from its muscles to the ground through work.

Energy exchange across an organism's outer surface—like its skin, or in the case of plants, etc., its body wall—is often more significant than energy exchanged via taking in or discharging chemical potential energy or via work. There are four physical processes by which organisms exchange energy with their environment across their outer surfaces—conduction, convection, radiation, and evaporation—and many animals use a combination of these to manage their energy budgets through each twenty-four-hour day-night cycle.

Transfer of energy by *conduction* can be explained by the fact that thermal energy naturally flows from zones of higher temperature to zones of lower temperature; it seeks to distribute itself evenly throughout an object or between adjacent objects. On hot days, dogs splay themselves down with their bellies touching cool stone floors. They are using conduction to transfer thermal energy from their bodies to the cool substrate.

*Convection* is the transfer of thermal energy by the bulk flow of matter, like water or air. Picture a hot animal, perhaps a buffalo, standing in cooler air: thermal energy will naturally move via conduction from the animal to the air immediately adjacent to its skin. As the temperature of this air increases, it forms an invisible "bubble" of warm air around the animal, and because warm air rises, it starts to move up and away from the animal, carrying thermal energy with it.

*Radiation* refers to the transfer of radiant energy, which is energy that travels at the speed of light, like the energy from the sun. But it's not just obviously "hot" objects that radiate energy; all objects (including living organisms) do, and the amount and type of radiant energy they emit is determined by factors like their size, shape, surface color, and surface temperature. Cool surfaces emit less radiant energy, so a Polar Bear, for example, can slow the loss of its internal energy via radiation by keeping its outer surface cool. It does this by having a layer of highly insulating underfur, which slows the conduction of energy out of its body, thus not raising the temperature of the outer layer of its fur.

The fourth—and often very important—means of energy transfer out of an organism's outer surface is *evaporation*, which relies on some of the special characteristics of water.

Water requires the input of a lot of energy to change from its liquid to its gaseous form, water vapor. When you sweat, you push water molecules out onto your hot skin. These absorb enough thermal energy to become water vapor, which floats away, taking large amounts of energy with it. This lowers your body temperature. Many animals speed up the process of energy transfer out of their bodies by panting; they expose the wet surfaces of their mouth interior and tongue to the air and then move air across these surfaces, increasing the rate at which water vapor is carried away.

Water is, of course, also an incredibly versatile solvent. Almost all organisms need to maintain a certain amount of water in their bodies to stay alive; they must do a dance of survival very similar to the obligatory dance with energy, and because of the important role that water plays in maintaining organisms' energy balance, these are linked dances. Organisms must often perform tricky balancing acts to simultaneously keep the correct amounts of energy and water in their bodies.

––––

Scientists have been closely observing Southern Yellow-billed Hornbills in a Kalahari study area since 2008. They've erected about forty hornbill nest boxes in the study area, substitutes for the natural tree cavities, to standardize nesting conditions and make nest monitoring easier.

Their research is producing remarkable—and shocking—results. In the ten years between 2008 and 2019, hornbill breeding success collapsed. Nestbox occupancy went down from 52 percent to 12 percent, success of the nests that were occupied from 58 percent to 17 percent, and the average number of fledglings produced per breeding attempt from 1.1 to 0.4. In short, the birds have gone from breeding at a decent rate in 2008 to producing almost no young in the early 2020s.

Why the stunningly rapid collapse? Once the female of the hornbill pair is cemented into the nest hole, the male needs to spend almost all of his day finding and transporting food for them both. Finding food is labor- and time-intensive work! But he must also constantly manage his internal thermal energy budget. When the sun comes up, he's immediately impacted by its rays, a significant source of radiant energy, much of which is absorbed by his body surface and becomes thermal energy inside him. He receives radiant energy not only directly from the sun but also indirectly from all the surfaces around him; it reflects toward him off the sand, off the grass and leaves, and off the base of any clouds that might be floating above. The more he moves his muscles to fly and hop in search of food, the more he transforms chemical potential energy into thermal energy inside his body. When he stands on the ground, energy passes into his feet by conduction from the hot sand, and the sand, of course, is also generating masses of high-temperature air, which rise up against him and transfer more thermal energy into him.

He can reduce the amount of energy passing into his body by staying in the shade; this cuts off most of the direct radiant energy from the sun. He can also stay still and not use his muscles, thus damping down an important internal source of thermal energy. But changing where he sits and what he does changes very little about his exposure to the thermal energy in the air. As the day unfolds, the air temperature rises, and when it rises above his body temperature, then thermal energy naturally, passively, begins to move from the air into his body by conduction.

When the male Yellow-bill's internal thermal energy starts to reach dangerously high levels, his last option for keeping his body temperature within limits is evaporation. He can actively transfer energy out of his body by turning liquid water into water vapor on the inside surface of his mouth, and he does this by panting.

The higher the daytime air temperature soars and the longer it remains high, the longer he must sit in the shade and pant, and the less time he has available to find food for himself, his mate, and their young. Hornbill researchers found that males were likely to engage in "heat dissipation behavior" when the air temperature went above 34.5 degrees Celsius (94.1°F), and that Hornbill breeding performance decreased in proportion to the number of days where the temperature exceeded 34.5 degrees Celsius. They also found that if the average maximum temperature during the breeding attempt was greater than 35.7 degrees Celsius (96.3°F), then no young survived. Average maximum daytime temperatures have been rising in the Kalahari. Nowadays a male Hornbill simply doesn't have enough foraging time in the day to keep his walled-in female well fed. When the young hatch and she can break out of the cavity to help him out, she faces the same temperature-management problem. Even with both adults working together, they don't have enough active hours in the day to find enough food for their babies. Climate predictions for the hornbill study area are that daytime maximum temperatures will exceed 35.7 degrees Celsius for every day of the breeding season by 2027: it will then be impossible for the birds to produce young, and the species will die out from the region.

The impact of the slightly higher average daytime air temperatures now being experienced in many arid areas is not dramatic and immediately noticeable. Birds don't fall out of the sky en masse. They just fail to reproduce and quietly fade away. Yellow-billed Hornbills can live for up to ten years, so it might be a long time before the absence of adults is noticed and their role in the ecosystem is no longer filled, but their species is just as surely doomed in the Kalahari as if a colossal heat wave, a week of twenty degrees above normal, were to kill them suddenly. All around the world, bird species are vanishing from hot, dry areas where they were recently common. Avian vacuums are forming in the hearts of arid zones.

———

Atmospheric temperatures are rising because humans have recently—within the last 250 years or so—begun emitting huge volumes of thermal-energy-trapping gases into the air, raising their concentration in the atmosphere. Some of this gas is subsequently absorbed by and held within the Earth's carbon sinks, like growing vegetation, the oceans, and certain types of rock that can chemically react with the gases and turn them into solids.

The most voluminous and influential of these gases is carbon dioxide. Most of the human-emitted (aka anthropogenic) carbon dioxide entering the atmosphere comes from burning fossil fuels, like coal, oil, and gas.

The sun is constantly sending energy toward Earth via solar radiation. Some of this is reflected off the outer atmosphere, but most of it passes into the atmosphere as visible light (shortwave radiation). Some of this energy is absorbed by Earth's surface, becoming thermal energy. Most of it, however, is reflected back toward outer space in a somewhat changed form, as infrared (longwave) radiation.

Different types of molecules interact with different wavelengths of radiant energy in different ways. Shortwave radiation coming from the sun passes directly through most of the gas molecules that make up the atmosphere as if they don't exist. But infrared radiation reflecting back from the Earth interacts strongly with some gases, including $CO_2$. When infrared rays hit $CO_2$ molecules, they transfer some of their energy to these molecules, causing them to vibrate faster and radiate energy out in all directions, including back down toward Earth's surface. This delays the passage of energy back into outer space, retaining it in the atmosphere and raising the atmospheric temperature.

$CO_2$ isn't the only naturally occurring gas that holds energy in the atmosphere. Methane, water vapor, nitrous oxide, and some others do, too. The fact that these gases are in the air is not in itself a bad thing. In fact, they make life as we know it possible; without them the biosphere would be so extraordinarily cold that the species currently living on Earth could not survive.

The problem of contemporary climate change arises because humans are pushing the concentrations of these gases higher than they've been in millions of years—higher than they've been during the entire time that many living species evolved in. We're also pushing up these gas concentrations at faster rates than they've ever risen at before, which is perhaps even more consequential for life on our planet.

The range of environmental temperatures that a particular species can tolerate is sometimes called its thermal envelope. Put differently, the upper and lower temperatures between which a species can survive define its thermal niche. If a species' environmental temperature goes above or below its limits, it has three options: it can adapt, move, or die.

Adaptation can be behavioral or physical; as we've seen, Yellow-billed Hornbills can survive short periods in ambient temperatures above the upper limit of their normal thermal envelope by changing their behavior—by sitting still in the shade instead of actively looking for food. Such behavioral adaptation can develop very quickly because it doesn't necessarily require a species to evolve genetically, but it has its limits, as we've also seen.

Adapting physically—for example, by developing a more efficient panting mechanism that would reduce the amount of time a bird needed to pant to keep its body temperature within safe limits—usually requires a species to evolve genetically. This takes generations, and depending on the species concerned, generations can be a short space of time or thousands of years.

Evolution doesn't operate at the individual organism level; it operates at the group level. In other words, we don't speak of an individual evolving but rather of populations or species evolving. Every organism in a population is slightly different from its parents and the other individuals in that population, partly because of random genetic mutations (changes) that occur during reproduction; it doesn't inherit fully perfect copies of its parents' genes. Sometimes these mutations make an individual fitter (better able to survive and reproduce) than its ancestors and contemporaries; sometimes they make it less fit. As Charles Darwin's *On the Origin of Species* famously explains, less-fit individuals die earlier and produce fewer offspring than fitter individuals. Over time, fitter individuals and their genes come to dominate, and thus the overall characteristics of the population or species changes; it evolves.

The ability of a species to evolve in response to climate or other environmental changes depends a lot on its evolutionary history. For example, did its ancestors evolve adaptations to similar environmental changes that occurred in the past, adaptations that are still at least partially encoded in its genes? Is it a species with a short generation time that can evolve rapidly?

Organisms can have fundamentally different ways of living with and dealing with thermal energy. Some are thermoconformers—they stay at the same temperature as their environment, like small fish, which tend to have the same internal temperature as the water they are swimming in—and some are not. Some are endothermic—they produce most of their own thermal energy internally, like mammals—while others are exothermic, like most reptiles, who gain most of their thermal energy from the outside.

These different relationships with thermal energy have a huge influence on a species' ecological niche, how large that niche is, and how flexible the species is in it. Some species are highly specialized: they have evolved to survive optimally in a very narrow ecological niche, under very particular circumstances. Others are generalists, able to survive well, if not optimally, in a wide range of environments and circumstances.

Some species have evolved to live in extreme climates, like deserts. They can tolerate very high ambient temperatures, develop efficient methods of offloading excess thermal energy, and retain scarce water very well—but there are points beyond which that even very well adapted species cannot go. There are limits to life. Some of the basic proteins in the body just cannot retain their structure and function above certain temperatures. They will fall apart, and the organism will die, no matter what. There is a certain minimum amount of liquid water that cells need to stay alive.

Some species that can't adapt to local climate shifts can move to regions where the climate suits them. If you're a mobile animal—especially if you can fly, like birds and many insects—this is theoretically easy. It's a different story for many sessile species such as terrestrial plants, whose individual members can't relocate themselves during their lives. With these, movement occurs between generations, for example seeds that separate from a parent plant and roll across the ground or are carried by the wind to their place of germination. The distance seeds can travel is often very short, and it takes many species years to move just a few feet.

There are also barriers to movement, habitats that species just cannot cross. For some species something as trivial-seeming as a narrow strip of a different type of soil can prevent movement. Fish can't just haul themselves out of water and take a short stroll over a small hill to colonize a new river.

And many species will die out unless they move together with one or more other species. As temperatures rise, many species of butterflies in California will have to move up in altitude or northward to remain within the range of temperatures they're adapted to. Although they can fly long distances, many butterfly species are heavily limited because each has evolved a tight relationship with a particular species of food plant. They must lay their eggs on a particular plant, on which its caterpillars have evolved to feed over millennia. If a butterfly species' food plant can't move upslope or farther north—because it can't project its seeds far enough, or grow in a different soil type—then the butterfly's ability to fly is meaningless.

But some butterflies are winning as the California climate heats up and humans change the environment. They are adapting to new food plants growing at higher altitudes. Some are finding refuge in urban gardens, which, because they are often shaded and watered, have cooler microclimates than the natural areas around them.

---

Joseph Grinnell and his co-workers didn't only observe and record birds during their surveys, but also mammals, and one of the most common mammals they found in the Mojave Desert was the White-throated Woodrat, a small rodent.

This species hasn't evolved much in hundreds of thousands of years. You could call it primitive or "ancestral," to be more polite. Its kidneys aren't very good at conserving water compared to more evolved, more "modern" rodents. It barely drinks and gets nearly all its body water from its food. So it has very limited ability to use evaporation to offload excess thermal energy.

On the face of it, the White-throated Woodrat would be a prime candidate for extirpation as the Mojave's temperatures go up. It should perhaps have disappeared already, along with so many birds. But recent surveys show its population doing well right across the desert. In fact, almost all the other small mammals in the Mojave are doing just as well as they were during Grinnell's surveys a century or so ago.

It turns out that this has nothing to do with these mammals' physiology, whether they're primitive or not, or how fast they're able to evolve. Their success can be explained purely by their behavior, which apparently has not changed since their species came into being. Nearly all the small mammals of the Mojave are nocturnal, active at night, when it's cool. During the day they sleep in underground burrows, insulated from the topside by several feet of soil. Thermal energy takes so long to travel through the soil that daytime temperature rises barely register deeper than a few inches. They spend the dangerous time in a quiet, dark, constant-temperature environment.

Some lizards aren't so lucky. At first glance they seem well-suited to live in deserts. They gain thermal energy from their environments, by sunning themselves or just hanging out in warm places. They don't need to break down food to generate metabolic thermal energy to maintain their body temperatures, as mammals must. A lizard thus needs to eat far less food than a mammal of equivalent size, a real advantage in the desert, where insects are relatively scarce.

Yet lizards have their thermal limits; if a lizard's body temperature goes too high, it will die, and, unlike mammals, lizards can't sweat to offload excess thermal energy. When the air gets too hot, all a lizard can do is stay still, out of the sun, in the coolest environment possible, and that usually means underground. But, just as for the Yellow-billed Hornbill, the time it spends staying cool is time that it can't be searching for food. A 2010 study found that some Mexican lizard species had already disappeared from the hottest parts of their range, and that perhaps 20 percent of lizard species would be extinct by 2080 because of global heating.

Few animals seem more vulnerable to overheating and drying out than amphibians, with their typically moist, thin skins. Many species of salamanders—amphibians that look superficially like lizards—live across North America, usually in moist habitats like damp forest floors alongside the streams in which they lay their eggs. But many of their habitats are changing fast. Forest streams that used to run all year round are drying up for months every year, and forest floors are getting warmer and drying out, too. Because salamanders are small and exothermic, their internal temperature rises and falls with their immediate environment's temperature. As it gets warmer, their metabolism speeds up, and they have to eat more food to maintain their body weight—but the insects they feed on are dying out because of the ever-longer dry periods.

Given this information, you might think that many salamanders are doomed, but they have a neat evolutionary trick. A salamander can respond to increased temperatures and drought by bowing out of the energy-and-water dance almost entirely: when it gets hot and dry, glands on its skin can exude a special slime that coats its whole body and dries into a strong, waterproof covering. It can burrow under the surface and, cocooned, almost completely shut its metabolism down. Its heart almost stops beating, and most of its internal systems simply stop working. In this state, salamanders don't need to exchange water or nutrients with their environment. They don't need to eat or drink, and they don't excrete waste. They can exist like this for years, until enough rainfall arrives, the streams fill up, and their insect food returns. When this happens they emerge from aestivation, eat like crazy, and lay eggs as fast as possible before drought returns.

In the same areas, fish are being challenged. Because they are also exothermic (cold-blooded), they need to eat more food to sustain themselves as their streams heat up. As a general rule, every 10-degree-Celsius increase in water temperature doubles fish metabolic rate. But there isn't enough food in some streams for hotter fish to find enough food; some are starving and their species are disappearing from rivers. A few fish species are evolving smaller bodies and slower growth rates as their water heats up. Being small requires relatively less food.

—

Individual plants have a problem that most animals don't face: once they've put down roots, they can't relocate. They have to deal with whatever the world throws at them, including the effects of climate change. Like other living things, plants need water, not just to fill their cells but to transport nutrients upward from their roots. They need water in the photosynthetic reactions in their leaves, where they use the energy from sunlight to combine it with carbon dioxide to build carbohydrate molecules, the basis for food chains everywhere. Plants let air—containing carbon dioxide—into their leaves through small controllable apertures on the leaf surface called stomata. Unless the stomata are open, no $CO_2$ can enter, and photosynthesis can't happen. Water in the plants also evaporates outward from open stomata, so when a plant's stomata are open, it loses water.

Desert plants have adapted to hold on to scarce water. They have thick skins, literally—waxy coverings, hard leaves, and many other adaptations to keep water in. They also have few stomata, which they control carefully to let in just enough $CO_2$ to survive and grow without losing too much water. (The little $CO_2$ they let in explains why desert plants typically grow very slowly.)

The Joshua Tree is the icon of the Mojave. It's a very good desert plant; slow growing, with a succulent trunk to store water and hard, tough leaves with few stomata. It can survive extreme temperatures. But its days in the Mojave are numbered.

Seedling Joshua Trees are too small to store much water. Their leaves are not particularly tough compared to adult leaves. They need an uninterrupted sequence of mild, relatively wet years to become mature enough to survive years that are hotter and drier than normal. In the lower-lying, hotter, and drier parts of the Mojave Desert, young Joshua Trees are already impossible to find.

Heat and drought also mean that once-rare fires are becoming common in the Mojave. The Joshua Tree has no defense against fires. They tear across the land, killing thousands of plants at a time.

Joshua Trees, which have lived in the desert for 2.5 million years, can't keep up with human-caused climate breakdown. A recent study predicts that by 2100, at least 80 percent of the Joshua Tree National Park will be too hot and dry for its namesake plant.

It's not only desert species that are being forced to move or die out, of course. Every species has its thermal envelope, and climate breakdown changes temperatures worldwide—mostly forcing them up, but sometimes down, too. Species everywhere are being pushed outside their thermal limits.

Ecologists often talk about a particular types of habitat or local ecosystem made up of definable, established communities of species that have evolved together into a more or less definable unit, like a certain type of forest, grassland, or wetland. It's already clear that climate breakdown will change them, take them apart, and force new combinations of species, habitats, and types of ecosystems to come into being. As the climate changes, not all species will be able to move with their climate envelopes; they'll die out. But some species will colonize habitats, areas, and communities that they have never lived in before. Climate change is scrambling natural communities and creating new combinations of species—new ecosystems—almost everywhere we look.

# Plagues & Diseases

There's a promise of change in the air on this bright morning in April 2021 in the forests of northern Maine, up near the Canadian border. The winter is slowly breaking, and the snow is patchily turning to slush. A few early wildflower bulbs are beginning to send sprouts up through the soil toward the light, and soon the migrant songbirds will arrive from the south. With their mouths wide open and their tongues shaking, they'll fill the greening trees with fresh notes, calling the summer closer.

But that will be in a couple of weeks. Here, now, is a young Moose, not quite a year old. She shuffles unsteadily toward a small birch tree. Desperately hungry, shivering, she gnaws on the tree's leafless branches, occasionally resting her body against its rough trunk. She seems at a loss. Her bones show their elegant forms through her skin; gently curved ribs, knobby knees, large shoulder blades, even the contours of her long skull, because her fur has thinned out and disappeared from large areas of her skin, some of which is turning white. Some parts of her body are covered with patches of glossy gray bumps, like dense sheets of unripe grapes that have been stuck on with glue.

She is what people have begun calling a ghost moose, one of many that now wander these parts. Within days she will falter to the ground, lower her head onto the cool, rotting leaves of the forest floor, and without ceremony breathe out her last.

We can begin the story of ghost moose like her half a century ago, with a species of small brown moth. The Eastern Spruce Budworm is a native moth that evolved in the conifer-dominated forests around the Great Lakes of North America. It is named after its inch-long brown caterpillar (the budworm), because this life stage has the most obvious effects on trees; it eats the needle-like leaves of fir and spruce species.

In most years Spruce Budworm exists in low numbers in the forests, feeding lightly on widely scattered trees. But every few decades, no one really knows why, its populations begin to grow exponentially, rapidly reaching thousands of caterpillars per tree. The caterpillars hatch in spring, begin chewing leaves, and by midsummer thousands or even millions of continuous acres of forest can be turning red, the color of dead needles. The larvae metamorphose into flying moths that emerge in late July; these can be swept up in their trillions by wind and deposited up to several hundred miles away. A Spruce Budworm outbreak is like a fire, consuming ever-greater acreages every year it lasts.

In the early 1970s Spruce Budworm numbers rose rapidly across eastern Canada, with infestations spreading south into the northeastern United States. Entire landscapes turned red; at its peak the outbreak covered an almost unimaginable 136 million acres, by far the largest area ever recorded. Although the caterpillar can feed on a few species of conifer and is named after spruce, it strongly prefers Balsam Fir, the most common tree in Maine. By the early 1980s it had spread across more than half of the state. Tens of millions of trees died.

Balsam Fir and spruce wood is valuable, so loggers rushed in to salvage the newly killed timber and haul it to sawmills before it could rot. They built thousands of miles of new logging roads into wilderness areas to clear-cut millions of acres—before the Spruce Budworm outbreak, logging had been smaller scale and more selective in Maine.

Because Balsam Fir formed the superstructure of these forests, its rapid removal completely changed the ecosystem. Almost overnight the forest landscape became open and light, creating new space for fast-growing, light-loving species like birches, maples, various shrubs, and willows, which moved in en masse, radically reshaping the landscape and ecosystems. Regenerating forests of small trees provide ideal foods for Moose, and Maine's massive new Moose pastures lured large numbers of these mammals from their traditional haunts north of the Canadian border and stimulated the fastest increase in Moose populations in recorded history.

The budworm outbreak died down in the mid-1980s, but by then Moose were common across Maine, New Hampshire, and nearby states, and by the 1990s tens of thousands of animals inhabited the region. The Moose is the largest species in the deer family, and large males can reach almost seven feet at the shoulder and weigh fifteen hundred pounds. Their size means they can open up pockets of habitat for species that thrive on disturbance, not only by eating but by trampling vegetation, thus diversifying the forest.

Although hunters came to shoot them, the changed habitat was so good for Moose that their population stayed high. In certain places they became more common than the ubiquitous White-tailed Deer, which did better in the pre-outbreak habitat. Moose soon became common in the landscape, as if it had always been this way.

But this state of affairs didn't persist. Around 2010 another newcomer, a tiny tick, would emerge on the scene. In a few short years it would radically change the fortunes of Moose in the far northeastern United States.

---

Ticks are not insects, as some people believe, but arachnids—members of the same taxonomic class as spiders. Fifteen tick species are found in Maine, and they survive by sucking larger animals' blood. Most of these types of tick have multiple host species during their life cycle—they live on different animals at different stages of life—but the Winter Tick has only one, its favored host being Moose.

The arrival of spring triggers adult female Winter Ticks to drop from their host animals to the ground where they each lay up to three or even four thousand eggs in leaf litter. The tiny larvae—less than one millimeter long, about the size of poppy seeds—hatch in late summer, and then crawl up into low vegetation. They cluster in groups at the tips of small branches or on grass fronds, sometimes in masses of hundreds or thousands, and begin to "quest" for a host. Larval Winter Ticks can pick up vibrations from walking animals, and even detect carbon dioxide in animal breath up to twenty yards away. When a tick detects a large mammal approaching, it reaches its forelimbs out into the air, and if the animal brushes against it, it hooks on to the animal's fur, often bringing its questing companions with it; the parasites interlock their limbs, forming clumps and chains that transfer together from vegetation to host.

Once larval Winter Ticks have successfully hitched on to a Moose, they burrow into its fur, penetrate its skin with their mouthparts, and begin to suck its blood. Over the next few weeks, they metamorphose into larger nymphs—about a quarter of an inch long—and then a month or two later transform into adults, which can grow much larger: blood-engorged female Winter Ticks can expand to the size of a grape (and the larger they are, the more eggs they produce).

———

Water has many remarkable attributes, one being that it can exist in gas (water vapor), liquid, and solid (ice) forms within the normal range of temperatures experienced in the biosphere. As a liquid, water molecules move around each other, held together relatively loosely by positive and negative electrical charges that attract the molecules to each other. As water loses thermal energy and its temperature drops, its molecules start moving around less, and it becomes denser, more compact.

But at about 4 degrees Celsius (39°F), small numbers of its molecules begin to join up into hexagonal ice crystals. Water molecules in ice are held slightly farther apart than when in a liquid state because of the nature of their chemical bonds, so as ice begins to freeze it becomes less dense and takes up more space. By the time water reaches the freezing point, all of its molecules are arranged in a regular crystalline structure and it takes up about 9 percent more space than when it was fully liquid.

Just as this expansion can burst a bottle of water placed in a freezer, it can burst and break the internal organs of organisms, killing them. Because they have such low mass, they have very low thermal inertia; they rapidly gain and lose thermal energy from and to their environment, which effectively means that their internal body temperature is determined by their surroundings.

As the air cools, insects and arachnids lose thermal energy, their body temperature drops, they stop moving, and, depending on how cold their immediate environment gets and which sort of organism they are, the water in their body fluids crystallizes into ice, destroys their internal structures, and kills them.

Many tiny species can survive cold winter months in a state called diapause, which is like suspended animation; they find places that won't get quite cold enough to kill them and shut their metabolisms down until spring raises their environmental temperature again, bringing them "back to life." Some species that have evolved in low-temperature climates contain antifreeze chemicals, which suppress and control the formation of ice within their bodies even at extremely low temperatures, protecting their internal organs from damage.

Winter Ticks, despite their name, don't have sophisticated anti-icing chemicals. When a Winter Tick is on a living host—a source of thermal energy—it can easily survive extremely cold weather, but without a host, it will die during the first prolonged ice and snow of winter.

Increased numbers of Moose provide more opportunities for questing larval Winter Ticks to find a lifesaving host for the cold season. But it's not only the numbers of Moose that determine a young Winter Tick's chance of surviving until the following spring; it's the length of time it can spend questing in the fall before the first deadly cold descends. The longer it can hang out in wait for a host, the more likely it is to find one.

Rising temperatures mean that Maine winters are now setting in about two weeks later than they did in the year 2000. This extends the tick questing period by about 25 percent, significantly increasing the chance that they'll hitch on to a Moose, and that they'll produce yet more young ticks the following summer.

So Moose are in a direct feedback loop, a vicious cycle, with their parasite, and adding higher temperatures to the mix is driving tick populations up at an even faster rate.

Most wild mammal species in the Maine woods, including White-tailed Deer, groom themselves regularly, pulling ticks from their skin. But Moose don't habitually do this, possibly because they evolved in cold areas where ticks have not been a big problem. Today there are orders of magnitude more Winter Ticks in the forest than there used to be, and Moose have few tools to combat them. A recent study in nearby Vermont found that the average Moose in that state carries 47,000 ticks.

———

Our young Moose was born in early June, just as the summer was hitting its stride. She was a singleton, even though Moose often had twins in the past. At first she grew well, but in fall she began picking up Winter Ticks, first hundreds, then thousands, then tens of thousands. As winter progressed they matured and grew larger, drawing more and more blood and nutrients from her body. Her diet of twigs was not enough to maintain her body weight. Her fur began falling out as she rubbed against trees in a vain attempt to rid herself of ticks. Her abraded skin began turning white. Her muscles started wasting away as her body used their protein to replace the blood sucked out by ticks.

By the end of winter, the female Winter Ticks she was carrying had ballooned to the size of grapes. They were bunched up against each other, bumpy sheets of parasites replacing fur that should have prevented her losing thermal energy to the surrounding air. But it wasn't cold that killed her in the end; massive weight loss and acute anemia did. The ticks exsanguinated her.

More than half of Moose calves born in Maine now die before their first birthdays, a far higher number than before Winter Ticks ran riot. Adult Moose are large enough to survive all but the most extreme infestations, though they can be severely weakened. Before Winter Ticks became common, young female Moose typically grew large and mature enough to become pregnant by their second autumn, at eighteen months of age. Now it takes a year longer, and most females only give birth every second year, failing to recover from the combined effects of pregnancy, lactation, and Winter Tick infestation to produce offspring annually as they used to. Twins are becoming rare. Despite increased death rates and declining health of Moose, there are still enough animals to sustain massive numbers of Winter Ticks in the northern woods. Biologists believe that the only sure way to restore Moose health is to radically reduce the number of ticks—by radically reducing the number of Moose. Hunters will have to shoot maybe half the animals to reduce their density enough to break the cycle of increasingly successful tick reproduction. Moose are suffering. They're failing to breed as they used to. A small temperature increase has helped to rapidly degrade the health and reproduction of the forest's largest animal for the indefinite future. We don't know the long-term ecological effects of this yet, but given the Moose's role in creating habitat for many other disturbance-dependent species, these may be significant.

———

Not long ago, around the time that European people began colonizing eastern North America, a unique ecosystem occupied about ten million acres along the coast between North Carolina and Nova Scotia. The Pine Barrens—a strange system by North American standards—has nutrient-poor, acidic, sandy soils, an ideal substrate for the Pitch Pine, a midsize pine tree species. Pitch Pine dominates this landscape, making it appear from a distance as a monochromatic, evenly textured, dull-green forest. Walk between the trees, though, and you'll notice that they're widely spaced. Quite a bit of light reaches the ground, which is rich in special wildflowers, including orchids of various kinds.

The Pine Barrens' species thrive thanks to fire, which has strongly shaped this ecosystem. For millennia, every twenty years or so a lightning-triggered blaze would work its way across the Barrens, burning back vulnerable vegetation: small, old, and dead Pitch Pines, as well as fire-sensitive species like oaks, and much of the underbrush, but leaving large, healthy middle-aged Pitch Pines alive with wide, sunny spaces between them.

Pitch Pines produce fire-resistant cones that open up shortly after the flames have passed, scattering seeds on the sandy soil, where they sprout up along with a unique community of fire-adapted grasses and wildflowers. The best-situated of the young trees grow fastest and become the next generation of seed-scattering adults.

About ten thousand years ago, as the last glaciation retreated, the ancestors of the Lenape Native American people lived in the New Jersey Pine Barrens. They were not able to grow abundant crops in the poor soil, but they found game animals and also useful meadow plants, which they set regular fires to promote.

When Europeans colonists arrived, they also could not grow their typical crops here (hence the name "barrens") and found the Pitch Pine to be too small and soft to produce good timber. But the land was cheap and easy to build on, so some constructed homes and settlements. As more settlers moved in during the 1900s, fire became viewed as a serious threat to houses and human lives; suppressing and fighting it became one of the government's chief concerns.

Without fire, Pine Barren ecosystems change. Mature Pitch Pines senesce, while other trees, including shade-tolerant oaks, fill in the grassy spaces between them. Pitch Pine seeds don't germinate well unless they fall on naked, fire-cleared sand, so few young trees grow. The ecosystem becomes denser and literally darker, forcing many species of wildflowers, butterflies, and reptiles out— there simply isn't enough sunshine or space for them anymore.

Today about 90 percent of the historical Pine Barrens ecoregion has been settled and urbanized, and only 10 percent persists in something like its natural form. The last significant patches of Pine Barrens habitat are in southern New Jersey, on New York's Long Island, and in coastal Massachusetts. With so many houses near the last remaining patches, fire suppression remains popular despite its well-known ecological costs, and the habitat is still in decline. But it's not only a lack of fire that imperils the Pine Barrens. A tiny new threat is spreading rapidly through the trees, perhaps the final nail in the coffin for this special ecosystem.

---

It's midmorning in a patch of Pine Barrens habitat on Long Island, and the sun has just risen high enough that its rays can directly illuminate the bark of a Pitch Pine tree. Within minutes the bark's temperature begins to rise, stirring a tiny Southern Pine Beetle nestled within it into action.

The beetle is the size of a rice grain and this tree is the only home she has ever known. She hatched from an egg here and grew up inside it, in the dark, transforming from a maggoty white larva into a stout, stubby-legged, armored red-brown adult. But now it's time to leave, and she chews her way a few millimeters forward and for the first time out into the daylight, where she spreads her tiny wings and buzzes away.

This first flight will be her only flight, and she doesn't go far. A mere ten yards into her aerial adventure she flies smack into the trunk of another Pitch Pine, which she immediately begins to chew her way into. The tree has an answer to her insult, though; as soon as she makes it through the dry outer bark and reaches the living tissue just beneath, it sends a little stream of pitch out toward her, a deluge of sticky, pine-smelling sap designed to trap her and push her back out of the tunnel she's just drilled.

The Pitch Pine's defense works, overwhelming the beetle's struggling limbs, but although doomed, she has already landed a decisive blow in what will soon become a full-scale war between her kind and the tree: she's released aggregation pheromones, airborne chemicals that let other Southern Pine Beetles in the vicinity know that she's bored into a juicy Pitch Pine and that they should join her. As soon as nearby beetles detect the pheromones, they redirect their flights toward this tree, and as soon as they begin munching into it, they release more pheromones, which attract yet more airborne raiders to focus their attacks here.

For a few hours the tree appears to hold its own, smothering and ejecting dozens of beetles with streams of exuded pitch, which mushroom out of the bark and solidify like bits of whitish popcorn up and down the trunk. But by the end of the day over a thousand beetles have chewed their way into its phloem, a layer of living tissue under the bark, in which they begin to excavate long, sinuous tunnels—galleries to lay their eggs in.

Each Southern Pine Beetle does not arrive alone. Just behind its head, around the front of the thorax, is a structure called a mycangium. It's a bit like a collar formed on the insect's body shell, in which it carries traveling companions: fungal spores. As the female beetles lay eggs along their newly bored galleries, they simultaneously inoculate the tunnels with fungi, which begin to grow on the inside walls, drawing nutrients from the tree.

Within days—less than a week—the beetle eggs hatch, the larvae begin to feed on fungus, and the upper needles of the pine begin to fade from green to yellow: the sheer number of beetle tunnels in the phloem tissue has broken the flow of water and nutrients between the tree's crown and its roots.

It takes less than a month for the larvae to become adult beetles, and by the time they are ready to chew their way out of their brood tree, its crown and most of its leaves have turned orange-red and it is dead. By the end of the summer this story will have been repeated countless times across Long Island. Thousands of acres of Pitch Pines will have died, and next year thousands more will almost certainly meet the same fate. This isn't normal. Southern Pine Beetles were unknown on Long Island until 2014, when they first began killing trees here. Recent years' higher minimum winter temperatures have allowed the Southern Pine Beetle to thrive, just as they have the Winter Tick.

———

The Southern Pine Beetle has long been found across Central America and the southeastern United States, mostly in areas with a subtropical or warm temperate climate. It mostly occurs in low numbers, living in scattered small groups, just another wood-eating insect among many. There are usually not enough of them to kill trees outright—they'll invade a single limb or dispatch one or two drought-weakened trees in a stand.

But occasionally Southern Pine Beetle populations will explode. As early as the 1700s European colonists recorded large patches of trees being killed by them in states such as Tennessee, Georgia, and South Carolina, although these outbreaks didn't last long, and dissipated naturally.

Because the Southern Pine Beetle evolved in mild climates, it lacks tools to prevent internal freezing in cold weather. It can burrow into trees and use their insulating bark and wood as a buffer from winter cold, but this only protects it for short periods of freezing air temperatures.

Until about twenty years ago, normal minimum winter temperatures were low enough to wipe out virtually every Southern Pine Beetle that ventured north of central New Jersey, but since then winter minimums have risen by about 1.5 degrees Celsius across many parts of northeastern North America. The region still freezes, but not quite as deeply as before. Large numbers of the beetles that fly north often survive now.

And the beetle's behavior has changed in the north, making it more powerfully destructive of pine trees. In the older, warmer parts of its range, the Southern Pine Beetle reproduces all year round, but in the new, cold regions, it stops breeding in the winter. Adult beetles burrow deep into Pitch Pine trees, and their metabolism shuts down. When spring brings higher temperatures, this means that virtually all the Pine Beetles in a particular area emerge and go looking for new trees in which to lay their eggs at the same time.

Cold winters synchronize Southern Pine Beetle emergence. The first warm day of spring triggers a massive simultaneous emergence, which can kill huge areas of pine very quickly. (In warmer regions, where beetle emergence is not synchronized—when a constant low number of beetles emerges and hits new trees from day to day—the likelihood of a tree's defenses being over-whelmed are very low. The beetles might just infest part of a tree and damage it a little.)

Since 2014, Southern Pine Beetles have colonized almost all of Long Island and wiped out tens of thousands of acres of Pitch Pines, transforming the remaining Pine Barrens habitat there. In the winter of 2021–2022, researchers found them in Pitch Pine forests in New Hampshire and Maine, an almost unimaginable leap northward. Southern Pine Beetles can now survive in virtually the entire range of Pitch Pine; there is now nowhere for the tree to live free from its nemesis.

As the band of rising winter minimum temperatures moves ever closer to the Arctic, cold air will no longer prevent the Southern Pine Beetle from moving north. But low temperatures aren't its only obstacle. At the northern limit of the Pine Barrens ecosystem the habitat changes to one dominated by different types of pines that aren't good hosts for the beetle.

Ancestral pine trees evolved almost 200 million years ago, and about 120 million years ago they split into two distinct groups of species called the soft pines and the hard pines, because of the relative hardness of their woods. Soft pines have pale, whitish wood and two or five needles in their leaf bunches, and hard pines have darker, more yellowish wood and two or three needles in their leaf bunches.

The Southern Pine Beetle overwhelmingly favors hard pine species like the Pitch Pine. It has been recorded burrowing into soft pine species like the White Pine, but not successfully breeding in them. At the northern edge of the Pitch Pine range, which the beetle has now reached, there is a broad band of forest about two hundred miles wide with no hard pines in it. This will stop the insect's northward progress unless something "completely unexpected" happens, like a truckload of beetle-infested wood getting driven through, or an unusually strong hurricane barreling up the east coast of North America and carrying insects over the soft pine belt.

There's lots of food for Southern Pine Beetles beyond the band of soft pines. Tens of millions of acres of boreal forest filled with another hard pine species, Jack Pine, lie to the north. Red Pine, another hard pine, is common in parts of eastern and central Canada and grows around the Great Lakes. If the Southern Pine Beetle makes it to these hard pine regions and winter minimum temperatures rise just a little more, then this single insect could devastate some of the largest forests on Earth.

———

The Southern Pine Beetle and the Winter Tick aren't the only tiny species having an outsize impact on the ecosystems of northeastern North America. Armies of microbes and insects are overhauling the forest ecosystems of this region at extraordinary speed, radically reducing or removing tree species after tree species. Unlike the Southern Pine Beetle and the Winter Tick, which made their way to new areas under their own steam, many of them have been transported here from other continents by people.

In the early 1900s a parasitic fungus called Chestnut Blight was accidentally introduced to North America from its native range in East Asia, where it causes mild disease in Asian chestnut tree species. At that time, American Chestnuts were an important component of many forests—they provided food for, among other species, the now-extinct Passenger Pigeon—but they had no resistance to the fungus, and it steadily moved through the continent, killing almost every American Chestnut it contacted. By one estimate over 3.5 billion were killed by the fungus between its introduction and 2013, and the tree is now functionally extinct in North America—so rare that it no longer fulfills its former function in the ecosystem. Only a handful of mature, wild-grown American Chestnuts survive today, almost all isolated trees growing more than six miles away from other American Chestnuts, a distance too far for the fungus to travel.

Other consequential insects include the Emerald Ash Borer, a small iridescent-green beetle from Asia that was first identified in North America near Detroit in 2002. The genotype of the introduced beetles has been traced back to central China, and it may have come to the United States in packing-crate wood. About a third of an inch long, the beetle lays its eggs under the bark of ash trees, including Green, White, Black, and Blue Ash—all the North American species. When the larvae hatch, they start chewing tunnels through the living tissue of the tree, killing it within a year or two. The Emerald Ash Borer contains sophisticated anti-freeze chemicals and can survive in air temperatures down to minus 30 degrees Celsius (−22°F). Its larvae can survive high temperatures of up to 53 degrees Celsius (127°F), so it's barely limited by climate anywhere on the continent.

Ash trees usually make up less than 10 percent of North American forests—usually much less—so their death can go unnoticed by casual observers, but their absence affects many forest-floor plant species. They typically leaf out a week to ten days later than other forest trees, meaning that the forest floor beneath them receives more spring sunlight, a boon for the various species of short-flowering wildflowers known as spring ephemerals (including the famous *Trillium* species) that cluster densely around them. When ash trees die, spring ephemerals suffer: forests without ash trees are forests with far fewer flowers, and at the rate the Emerald Ash Borer is spreading, in less than twenty years this will be virtually every forest in North America.

Another ecologically damaging insect is the Hemlock Woolly Adelgid, a tiny bug—only about 1.5 millimeters long—first found in the eastern United States in the 1950s, when it was accidentally introduced from Japan. Young adelgids attach themselves to the bases of hemlocks' needle-like leaves and insert their mouthparts to suck on the trees' juices. They remain attached in place for their whole adult lives, depositing a small egg mass covered by a white, wooly-textured, waxen covering just before winter. Millions of adelgid bugs can infest a single tree, sucking it to death over a period of four to ten years. Adelgids can be killed off by very low winter temperatures, but as these have risen, the bugs have moved north. They are now found across the entire range of Carolina Hemlock and all but the northernmost parts of Eastern Hemlock territory.

Scientists predict that almost all hemlocks in adelgid-infested areas will be dead by 2040, with their loss being felt especially strongly by aquatic species. These dense, stately trees prefer to grow along streams, where they reflect and absorb solar radiation before it can hit the water. Northeastern streams warm up fast without their hemlocks, making their waters unsuitable for many species that evolved in them. Insects die, frogs can no longer breed, and trout vanish; the streams become pretty aquatic deserts, winding through the hills.

———

Imagine an ordinary midsize antelope species, like one you'd see running away from a lion in an African wildlife documentary. It's tan all over, with good running legs and small hooves. The males carry mid-length horns; the females are bareheaded. But this creature's nose is very odd: it bulges up all the way from its forehead to the tip of its snout, where it terminates in two absurdly large, forward-facing nostrils, somewhat as if the end of an elephant's trunk has been attached along the front of its face by an imaginative plastic surgeon. This is a Saiga, one of the strangest-looking antelopes on Earth.

The Saiga evolved about a million years ago, during the Pleistocene epoch. Before humans became civilized, it ranged in dense herds from the British Isles and across Europe to Central Asia, Siberia, and much of North America. Today it is found in the high-altitude shortgrass habitats of Central Asia; one subspecies lives in Mongolia, the other in Kazakhstan and Russia. It's a nomadic species, constantly on the move, traveling hundreds or thousands of miles back and forth across the plains.

The Saiga is the most significant grazer in its range, turning more vegetation into meat and nutrient-rich feces than any other animal. By cutting down the grass, it opens space for small mammals and birds to breed, and by churning the soil with its hooves and fertilizing it with its droppings, it aerates and revitalizes the soil. Live Saiga feed large numbers of predators, including wolves, eagles, and foxes, and dead Saiga sustain scavengers like bears, vultures, and flies.

The Saiga's bizarre oversize nose allows it to thrive in a challenging climate of hot, dry summers and freezing winters. In summer, when the plains are dry, it filters out the fine particles of dust constantly churned up by the herd's hooves. By pumping high-temperature blood into fine vessels on the wet inner surface of the nasal passages, the nose helps the Saiga offload excess thermal energy via evaporation with every breath out. In winter, hot blood in the Saiga's nasal passages warms freezing air on the in-breath, before it reaches the lungs.

In fall and winter male Saigas spar for the right to lead small harems of females, with whom they mate. In late winter the adult males move off to higher altitudes, while the females and juvenile males gather in vast calving herds in lower-altitude grasslands with somewhat warmer air. Then, in May, the females begin to give birth, two-thirds of them having twins and the remaining third singletons. Saiga babies have larger bodies and longer legs relative to their mothers than any other antelope species. The young must be able to keep up with the moving herd immediately after birth or risk being picked off by predators like wolves and golden eagles, which constantly shadow them.

The largest remaining Saiga population lives in the Betpak-Dala ("Hungry Steppe") region of central Kazakhstan, a gently rolling semiarid flatland covered with sagebrush and grasses.

In May 2015 the females of this region had gathered together as normal in a megaherd of over two hundred thousand animals. The air temperatures were a little warmer than normal, and there had been better-than-average rain, which stimulated strong grass growth. The first young were soon born, healthy and large, but then a strange affliction began moving through the animals.

As if a bad spell had been cast on the herd, thousands of females suddenly became weak. They lay down, lethargic and depressed, uninterested in moving with their herdmates or feeding. Soon they began frothing at the mouth, bleeding from their noses, and pushing out bloody diarrhea. Hungry young stood alongside their mothers, occasionally kneeling down to suck desperately at their teats. Within three or four hours of lying down, the females died, and within hours their nursing young had lain down and died, too.

Within five days the entire herd was dead. Not a single survivor was found. Conservationists surveying the scene from the air estimated 211,000 dead Saiga on the plains, 88 percent of the Betpak-Dala population. The surviving 12 percent were mostly males living in small groups distant from the calving herd; mysteriously, whatever had wiped out the calving herd didn't affect them at all.

It took years for scientists to figure out what killed Betpak-Dala's Saiga. They took soil samples, vegetation samples, even air samples, as well as tissue samples from the dying and dead animals. They diagnosed the cause of death as hemorrhagic septicemia—blood poisoning accompanied by bleeding, which can be set off by a large range of pathogens—but they found nothing strange or toxic in the soil, air, or forage that would have caused it.

After testing the Saiga tissue samples for common diseases and coming up with nothing—the animals appeared to be in excellent health before being struck down—they started looking for more obscure microbes and toxins. They eventually found extraordinarily high levels of a bacterium called *Pasteurella multocida* in every tissue sample from a dead or dying Saiga. *Pasteurella multocida* is commonly found in the nasal passages and on the tonsils of Saiga, one of the many harmless and seemingly inconsequential bacteria that make up the internal microbiomes of so many species, but for some reason it had multiplied out of control, spread throughout the Saiga's bodies, and poisoned their blood. The toxins generated by *Pasteurella multocida* were then passed to the nursing young via breast milk, which is why they died a few hours later than their mothers.

A fine-scale analysis of the weather in the Saiga calving region in the days leading up to the mass die-off revealed an anomaly. For ten days before the female deaths began, there had been double the normal amount of rainfall for that period, and the air had been three degrees Celsius warmer than normal. This combination drove air humidity consistently above 80 percent, a highly unusual situation in central Kazakhstan, where air is usually dry. Saiga nasal passages had thus also been consistently humid for ten days, ideal conditions for *Pasteurella multocida* to multiply rapidly, spread across the soft tissues inside the nose, and pass into the blood.

This was not a new microbe. Saiga may have been living with *Pasteurella multocida* in their bodies for as long as their species has existed. All it took was a few days of strange weather—the sort of weather that's becoming more common in many parts of the globe as air temperatures rise—for a formerly harmless symbiont to turn deadly and abruptly reduce the movement of nutrients and energy through an ecosystem covering hundreds of thousands of acres.

---

The Hawaiian Islands are some of the most isolated bits of land on Earth, just a few alluring tropical dots in the heart of the vast Pacific Ocean. They owe their existence to a rare "hot spot" just below Earth's crust, where superheated magma melts its way up through the sea floor to build slow-growing seamounts (underwater volcanoes) that eventually emerge above the waves in clouds of steam and smoke as newborn islands.

The Pacific Plate, the massive tectonic plate that underlies most of the Pacific Ocean, moves northwestward at the almost unimaginable rate of thirty-two miles per million years, but the Hawaiian hot spot remains fixed in place. As the plate moves, the hot spot melts new holes and disgorges more magma through Earth's crust in the "wake" of the seamounts or islands it's just produced, which is why the Hawaiian Islands are laid out in an age-ordered line. The oldest major island, Kaua'i, the farthest from the hot spot, is about 5.1 million years old.

Despite Hawaii's extreme distance from large landmasses, a good variety of terrestrial life forms have found their way to its islands, survived, and radiated into new species. There is only one type of native land mammal, the Hawaiian Hoary Bat—Hawaii is just too isolated for flightless land mammals to have floated there alive—but the islands have nurtured hundreds of types of colorful land snails and fascinating native birds such as the Hawaiian honeycreepers, an endemic group of small songbirds that has developed an unusually diverse array of forms there.

Many honeycreepers evolved long, thin bills to feed on nectar. Others evolved short, strong finchlike bills to tackle seeds. And some evolved to specialize on insects or snails. Honeycreepers come in a kaleidoscopic range of plumage colors, and most species don't immediately appear related to each other, but they are: the most recent research shows that all known species of Hawaiian honeycreeper evolved from a single colonizing species. This was likely a relative of today's rosefinches, small seed-eating songbirds, most of which have reddish plumage and are found in Central Asia.

It might seem strange that the Hawaiian honeycreepers' ancestral species came from so far away, but rosefinches are particularly prone to irruptions: sudden, long-distance movements in large numbers. At the slightest hint that their food sources are dwindling, rosefinches will flock together and fly off in a seemingly randomly chosen direction to seek richer feeding grounds. These trips sometimes take them halfway across continents.

Genetic evidence points to a group of hungry rosefinch ancestors heading out over the Pacific about five million years ago, perhaps blown by a storm, and somehow finding the stamina to keep themselves above the waves for thousands of miles until, with extraordinary luck, they ran into a young Kauaʻi, where, with yet more luck, they found enough of the right kind of food to survive and breed.

Over the next five million years this colonizing species split and evolved into at least fifty-six different species scattered across the archipelago. (This is how many honeycreeper species scientists think were alive when humans first arrived in Hawaii.)

But only seventeen species of honeycreeper survive today, and all but three are threatened. Despite Hawaii's seductive tropical paradise image, these are islands of extinction.

The first human settlers, Polynesians who arrived about a thousand years ago, overhunted many species. As more people from different places colonized the islands, they brought animals like rats and pigs with them, which went feral, killed native species, and wrecked their habitats. Colonists razed native forests to grow crops and graze domestic animals. All these factors drove many Hawaiian birds extinct, especially at lower altitudes where human populations were denser. Today, however, the most significant threat to native birds is avian malaria.

Hawaii has no native malaria-carrying mosquitoes, but the insects were introduced to the islands in 1826, arriving in water barrels on whaling ships. For about a hundred years they had no more than nuisance impact, but in the early 1900s European settlers, who missed the sound of birdsong in the lowlands, introduced colorful songbirds from distant countries, some of which apparently came with avian malaria parasites in their blood. Mosquitoes soon began to transfer these parasites from the introduced birds to the native species, and honeycreepers at low elevations started to die off in huge numbers. By the 1940s most lower-altitude honeycreepers were gone. Higher-elevation birds were not initially hit by malaria because their habitats were too cool for mosquitoes. But as Hawaii's air temperatures have risen in recent decades, mosquitoes have been able to bring death up the island's slopes.

In 2022 the Hawaiian state government and the U.S. Fish and Wildlife Service announced that four of the surviving honey-creeper species were in imminent danger of extinction because of avian malaria. Mosquitoes had reached the highest altitudes of Kaua'i, and nowhere on that island was safe from the disease anymore, they said. Only forty-five of the tiny, agile, insect-eating 'Akikiki remained in their last habitat on Kaua'i, and malaria was killing them so fast that they could be gone from the wild by mid-2023; by the time you read this book.

Another Kaua'i honeycreeper, the 'Akeke'e, a yellow-green bird with an asymmetric, crossed bill that it uses like scissors to cut open buds to find insects, was also in imminent danger of extinction from malaria, as were the Hook-billed Kiwikiu and the Crested 'Ākohekohe of the island of Maui.

Conservationists have taken a few of these honeycreepers into captivity, hoping to keep at least a few birds breeding in cages away from the disease. In May 2023 they released lab-bred mosquitoes infected with a sterility-inducing bacteria in bird habitat, trying to crash the mosquito population. Time will soon tell whether these last-ditch efforts have any chance of success.

---

About 2.7 million years ago, the Arctic region began to freeze, setting in motion the recurring cycles of freezing and thawing that marked the Pleistocene era. Some parts of the subsoil in the far north have remained continuously frozen since then, and other regions have periodically thawed, only to freeze again. This frozen subsoil is called permafrost, because during the entire time that our species has existed it has never thawed (unlike the sea ice and the soil surface, which have long done so every summer).

Many Arctic soils become soft and muddy when they melt. The tundra vegetation that grows on them accretes in layers, living plants growing out of and above their dead predecessors, building up ever-thicker layers of acidic, oxygen-poor peat. Peat and ice, combined, are excellent preservatives, and uncountable numbers of animals that died in the far north are sequestered just under the surface, often almost undecayed. Woolly mammoths, hairy rhinoceroses, and dozens of other types of long-extinct mammals are buried with their flesh intact (even edible!) after tens or hundreds of thousands of years.

Like animals alive today, these extinct beasts contained rich populations of microbes, including viruses and bacteria that cause disease. Now, with the Arctic warming at three to four times the global average rate, thousands of acres of permafrost are melting. Carcasses are emerging from the ground, along with their pathogens, which, because they have been out of circulation for tens or hundreds of thousands of years, are effectively new to animals and plants alive today.

Lab scientists have already revived disease-causing viruses from thirty-thousand-year-old permafrost, and soon hope to revive viruses that have been frozen for over a million years, but no pathogen that's recently emerged from the permafrost has yet been proved to cause serious disease; a 2016 anthrax outbreak among Russian reindeer that killed a reindeer herder's child has not conclusively been linked to thawing animals. But large-scale melting has been noted only since 2010, and only a tiny percentage of the permafrost has melted since then. The process has just begun.

The next major ecosystem-rearranging disease may be an old one, set free by warming.

Climate change doesn't only bring advantages for plagues and diseases. It can also suppress them by making climates inhospitable. But there's been very little evidence of this happening so far, and much more evidence for climate breakdown helping diseases to break out and spread.

And not every climate change–induced disease breakout becomes a long-term disaster. Since the Saiga antelope population crashed in central Kazakhstan in 2015, the unusually hot and humid weather conditions that triggered their massive die-off have not returned. In the following years the population has grown rapidly, and in 2022 almost half a million Saiga were counted in the Betpak-Dala population—many more than the precollapse total. Saigas' ability to bear large numbers of young, a product of having evolved in an unstable environment where drought is common and species must exploit good years to the maximum, has stood them in good stead.

Moose will probably persist in Maine, but as unhealthy animals. Without intense human intervention, the unique Pine Barrens habitats of the northeast are likely doomed, along with the ash and hemlock species of that region. Northeastern woodlands face ever-intensifying waves of pests and pathogens, each one functionally removing more tree species, changing their structure and composition, simplifying them and thinning them out—and perhaps making space for species moving up from the south.

Scientists are just beginning to research how climate breakdown can influence the breakout and spread of plagues and diseases across the planet. The little we already know indicates that almost all ecosystems can be rapidly thinned out or fundamentally transformed by them, and once their effects are prominent enough to be noticed, it's often too late to stop them. It's very likely that many emerging pests and pathogens are invisibly building up their numbers in new territories right now.

3

# Extreme Weather

The evening of Tuesday, September 19, 2017, is a gentle one in the El Yunque rainforest, a verdant blanket of trees that softens the Luquillo Mountains of northeastern Puerto Rico. It's warm, and there's almost no wind. Earlier in the day the sky was brilliantly clear, but now, as the sunlight begins to wane, clouds have moved in, and a tropical drizzle is falling. All around, countless million bright green leaves are shining, dripping, wet.

We don't know if the rainforest's most famous residents, a flock of about fifty-five Iguacas—Puerto Rican Parrots—have any idea what tonight has in store for them. Are they behaving differently from normal? The human conservationists who might be up here, in the forest, watching, have been evacuated.

Most likely the Iguacas are following their regular late-afternoon routine, which began around 4:30 P.M., when they roused from the hours of rest they take during the midday heat. As the day cools, they're flying between trees, looking for food around their habitual territory in a small valley along the Espíritu Santo River, which rushes and bubbles its way through El Yunque.

The forest's continuous leafy canopy lets very little light through, allowing only a few of the most shade-tolerant species of understory plants to grow, so it's surprisingly open, almost parklike, between the soaring trees. You can see through the forest for tens or even hundreds of yards. But even though the Iguaca are large and loud, they can be hard to observe; their rounded, short-tailed bodies are wrapped in almost completely green feathers that blend into the swells and sprays of foliage above.

Close up, the birds show dramatic, expressive faces, with a bright red strip of feathers across the forehead and a clean white patch of facial skin, like fresh-painted leather, around each lively gold-and-black eye. As with most parrots, Iguacas' feet are strong and handlike, supporting the birds as they crab-walk along branches or hang upside down, and are used to pull food toward their eager mouths. Their stubby fingerlike tongues help to manipulate wild fruits into the perfect place to be cut to pieces by their powerful, horn-colored beaks.

This evening, like all evenings, is a social occasion. The Iguacas communicate constantly via loud screeches, grating squawks, and whistles that project far through the greenery. They have special takeoff calls, to let each other know when they're leaving, and in-flight contact calls to stay together while speeding across the forest. They use warning calls when raptors are near and a wide range of other vocalizations to express affection, aggression, or fear and pass along information. Raucous ongoing conversation cements their lifelong monogamous pair bonds.

This flock of Iguacas that lives along the Espíritu Santo is particularly special, because about half their number—roughly twenty-five birds—have unbroken wild ancestry; they were born from lineages that have been living free on Puerto Rico for well over half a million years. The other half were born and raised in the cages of a captive breeding program, or are recently descended from such captive-raised birds. Although physically and genetically the same, the birds with purely wild ancestry and those from captive stock are culturally different. The wilder birds have a significantly larger, more diverse vocabulary.

The humans who watch and protect Puerto Rico's Iguacas can't yet understand the details of the different birds' languages, but they can distinguish them by ear right away: the wilder birds utter shorter notes in more complex phrases than their cage-raised flockmates.

On this September day there are four groups of Iguacas totaling about 650 individuals left on Earth. They are all in Puerto Rico, but only this El Yunque flock contains wild-talking birds, the last custodians of the old Iguaca language. The other three groups—two captive in conservation breeding programs and one free-flying but built up solely from captive-raised birds—speak only the impoverished tongue of the cages.

Maybe, like some animals, the Iguacas of the El Yunque flock can sense this evening's rapidly falling air pressure. Maybe they know what this means. Maybe the wild-talking birds are urgently trying to transmit ancient knowledge about the oncoming weather to their flockmates. Maybe they're dropping down to the forest floor, seeking out shelter in the hollows between large tree roots before it gets too dark to see. Or maybe everything seems normal to them and they're splitting up into their regular pairs and small family groups to find high, grip-sized branches to settle down on, gently dropping their heads forward or tucking them under their wings, closing their eyes, and going to sleep.

---

About 3.5 million years ago some large green short-tailed parrots belonging to the group we know today as Amazon parrots (genus *Amazona*) had made their way from the Central American mainland to a landmass now known as the island of Jamaica. This wasn't as hard as it might seem, because at that time large sections of a massive ridge called the Nicaraguan Rise, which runs between the mainland and Jamaica, were exposed above the waves, providing convenient stepping stones for parrots (and other creatures) to move along.

The ecosystems and climate of Jamaica were different from those of today. Although North and South America were slowly— very slowly!—moving toward each other, they were not yet joined, and the waters of the Pacific Ocean and the Caribbean Sea mingled in the space between them. The western Caribbean was washed through by powerful cool currents carrying nutrients from the deep Pacific that fed vast, productive underwater ecosystems and moderated the climate across the region.

But the Caribbean is characterized by complex and relatively rapid ongoing geological change, and the Rise was already sinking when the parrots moved across it. Within a few thousand years it completely disappeared underwater (it's about four thousand feet down today), marooning the birds on Jamaica, where they began to evolve into a new species, distinct from its mainland ancestor.

About 700,000 years later (2.8 million years ago) the Isthmus of Panama finally closed, creating a land bridge between North and South America and cutting the Caribbean off from the Pacific. This kicked off the Great American Biotic Interchange, in which terrestrial species from South America colonized North America and vice versa, changing land-based ecosystems on both continents. Animals such as tapirs, deer, horses, big cats, canids, rodents, and elephant-like gomphotheres expanded southward from North America. Marsupials, strange, massive ground sloths, terror birds, and huge armadillo-like glyptodonts and pampatheres went north from South America.

The closing of the isthmus also triggered the formation of the Gulf Stream, the fast-moving warm ocean current that arises in the Gulf of Mexico, travels up the east coast of North America, then moves eastward across the North Atlantic all the way to Europe, influencing the climate of half the Northern Hemisphere on the way. It radically changed Caribbean marine ecosystems by stopping nutrient-rich upwelling currents, triggering a mass extirpation of shellfish, corals, and other marine animals, leaving behind mostly species that preferred low-nutrient waters.

The parrots that had colonized Jamaica didn't just survive through these and many subsequent climatic and environmental shifts; they thrived. Their intelligence and communication skills helped them form strong cooperative pair bonds, family groups, and flocks. Their powerful wings could carry them to distant fruiting trees and better habitats if local ecosystems changed and no longer suited them. Over hundreds of thousands and then millions of years, they colonized other Caribbean islands, step by step. They evolved into new species on different islands, each with its own language, and each developed distinctive color marks (the Cuban Amazon has a white forehead and pink throat, for example) while retaining the overall green plumage that is typical of their group.

The youngest Caribbean species of Amazon parrot is the Iguaca, which evolved from birds that moved across the water from the island of Hispaniola and successfully colonized Puerto Rico about 670,000 years ago.

It would be a very long time before Iguacas encountered humans, who, it appears from archaeological records, settled the island only about four thousand years ago. These first inhabitants were Ortoiroid people, hunters and fishers originally from the Orinoco Valley region in present-day Venezuela, who fanned out across the Caribbean. The Ortoiroids were absorbed and superseded by successive waves of American Indian immigrants, each more technologically advanced than the last. Between the seventh and eleventh centuries the Taíno indigenous culture developed on Puerto Rico and other Caribbean islands. (It was the Taíno who named the parrots "Iguacas," an onomatopoeic derivation of their loud flight calls.)

Although there's little evidence that Taíno people deforested Puerto Rico to any great extent, they had an impact on the ecosystem and hunted many native land mammals into extinction. There were likely some tens of thousands of Taíno living on Puerto Rico when Christopher Columbus landed during his second transatlantic voyage in 1493 and claimed it for the Spanish crown. He stayed only two days, but in 1508 one of his lieutenants, Juan Ponce de León, established the first permanent European colonial settlement on the island. The colonists began mining gold and engaged in some limited farming. Early Spanish records describe Iguacas as being abundant across the island; there may have been half a million or more of them, even though Taíno people sometimes ate them and probably kept them as pets.

The colonists subjugated the Taíno and exposed them to European diseases to which the islanders had no immunity. They soon died in massive numbers. Then the colonists replaced them with indigenous slaves from elsewhere in the Caribbean and black slaves from Africa who intermarried with surviving Taíno, building a new creole culture together with their white overlords.

The Spanish crown paid little attention to Puerto Rico for the next two hundred years, and the colony languished economically, supporting only a small human population. Few humans meant abundant forest, and Iguacas remained common.

This changed in the 1700s, when the Bourbon rulers of Spain strongly encouraged immigration from Europe and elsewhere and vastly expanded agriculture. Trees fell left and right, and by the close of the century Puerto Rico had a population of 155,000 (including over 13,000 slaves) and was a significant exporter of cane sugar and coffee. This trend continued in the 1800s; coffee and sugar became even bigger industries. In 1876 the Spanish king proclaimed a few thousand acres of forest in the higher reaches of the Luquillo Mountains as a reserve. He wanted to prevent rival powers from building warships with the wood and threatening his growing colonies in the Caribbean. By the 1890s almost a million people lived on the island.

In 1899, after the Spanish-American War, Spain ceded Puerto Rico to the United States. The island's new rulers expanded agriculture even more, promoting animal ranching, fruit growing, and row crops. In the early 1920s only 1 percent of Puerto Rico remained as primary forest—the few thousand acres of the Luquillo forest reserve, which the Americans had kept under protection and would later name El Yunque. By 1940 this was the only place on Earth where Puerto Rican parrots survived. Some two thousand birds were estimated to still live here at that time, but they were in exponential decline. The cause eluded biologists, as the forest remained in good condition. By 1950 there were only about two hundred parrots left, with no sign of population stabilization.

Iguacas are strong fliers and can move many miles between habitats. They're not picky eaters and can live on fruits from many different types of trees. Even in the mid-1900s, when Puerto Rico probably had less tree cover than at any time in their species' existence, there were still several forested areas around the island with enough fruiting plants to feed groups of Iguacas year-round. But the parrots survived only in El Yunque.

The cause of their confinement to this single forest is their need for particular nest sites; they breed in hollow chambers inside tree trunks or large branches. Iguacas cannot carve these chambers out themselves—they need to find holes made by natural forces, like those formed in tree trunks when large, old branches break out, or excavated by other creatures, such as abandoned wood-pecker nests enlarged by decay. In technical terms, Iguacas are obligate secondary cavity nesters, and the cavities they need are not easily found in the young, healthy trees of regrowing secondary forests. In 1940 El Yunque was the only place they could find primary forest containing the gnarly, life-worn trees they needed to breed in. All the island's other forest patches were reemerging stands that had previously been cut down.

But it turns out that even this relatively large reserve with old wood and at least sixty-five types of Iguaca food plants could not sustain the parrot over the long term. In fact, had it not been for extraordinary, sustained human intervention, El Yunque would likely have been the species' doom. The Iguaca's confinement here turns out to be a vivid, sobering illustration of the insidious, many-fingered grasp that climate can hold and squeeze a species in, and how a combination of subtle, long-acting climatic forces and just one extreme weather event can snuff a species out.

———

In 1967 the Puerto Rican Parrot was among the very first species legally listed as "threatened with extinction" in terms of the US Endangered Species Preservation Act,§ which funded skilled

---

§  A predecessor of the well-known Endangered Species Act of 1973.

people, equipment, and facilities to save it. By the next year, the first of what would be many teams of biologists was assigned to El Yunque to conserve the birds. But as they fanned out into the field, they could only find twenty-four remaining parrots. Conservationists further found that less than a quarter of nests produced fledglings in an average season, and the flock was fading away. (It would bottom out at a mere thirteen birds in 1975.)

The specialists assigned to save the Iguacas soon realized that many nests failed because they were too wet. The Luquillo Mountains are the wettest place in Puerto Rico, with typical annual rainfall of 2,000 millimeters in their lower reaches, and 5,000 at their crests. The area occupied by the last wild parrot population, between 500 and 700 meters above sea level, gets an annual average of about 3,500 millimeters—sopping by any standards. The rain penetrated nest cavities, cooling eggs and causing fungus to grow. Chicks got soaked and died. El Yunque's nest cavities rotted into disrepair faster than those in dryer localities would have done, lasting only ten or fifteen years.

Conservationists repaired and waterproofed old cavities and began hanging artificial nests made from large PVC plastic drainage pipes covered in chunks of bark and wood in parrot territories, which the birds soon began moving into. Hatching rates went up, as did nestling survival—but the majority of Iguaca young still died in the nest or within days or weeks of fledging. The wild population barely grew.

Biologists determined that the more rain that falls while a baby Iguaca is in the nest, the greater the chances that it will become infested with botfly larvae, pale, slimy maggots that burrow head-first into baby birds' skin and then deeper into their flesh. Botfly maggots thrive when the air is more humid, and more rain means more humidity, for longer. (The new artificial nests made no difference. Even the most rainproof plastic nest doesn't keep air out.)

A competent conservationist can remove botfly larvae one by one by carefully gripping them with sharp-pointed tweezers and gently pulling them out, but the maggots can leave parrots' body tissues, like flight muscles, irreparably damaged. Botfly infestation survivors often become weak, uncoordinated fledglings, doomed to an early death.

Climatologists found that heat and humidity have been rising in El Yunque in recent decades not only because of overall global atmospheric warming but because deforestation in nearby parts of the island has increased the amount of heat radiated from the ground. Higher heat and humidity boost fungal and bacterial growth, which helps to explain why Iguaca biologists found that bacterial and fungal diseases infect about a third of El Yunque's wild-born parrot young while they are still in the nest.

Common among these maladies is *Aspergillus*, an often-deadly fungus that infects a bird's respiratory system. Other researchers analyzing (of all things) dust have found a reason why this fungus is so prevalent in the Iguacas of El Yunque: in recent decades, ever-increasing amounts of airborne dust laced with non-native bacterial and fungal spores—including significant amounts of *Aspergillus*—have been landing on Puerto Rico. Lab work has shown that most of these spores are not of nearby origin but from Africa, whipped up from degraded land and carried clear across the Atlantic Ocean by strong trade winds. The threats to the Iguacas of El Yunque have been magnified not just by local climatic conditions but also by desertification and deforestation (partly caused by climate change) on another continent.

Biologists also figured out that if heavy rain fell within three days of a young parrot leaving the nest for the first time, it would be far more likely to die than a parrot that fledged in a dry spell. New fledglings were naive to rain and did not shelter themselves well. They became soaked and struggled to thermoregulate, and disease set in. Their beginner flying skills were degraded by heavy, waterlogged feathers. They became easy targets for predators, even for ground-living carnivores like the introduced Small Indian Mongoose, which normally wouldn't be a threat to parrots.

In 1973 conservationists began removing eggs from the last Iguaca nests in El Yunque and hatching them in captivity. This had two benefits: it created a captive insurance population of the species, and it significantly increased the total number of Iguaca iguaca eggs produced. Iguacas, like many birds, tend to double-clutch, laying a second batch of eggs if their first one is lost—or, in this case, "stolen" by conservationists—early in incubation.

Some of the captive-hatched nestlings were raised by hand, but many were raised in the breeding facility by Hispaniolan Amazon Parrot foster parents. Iguacas could not be used as foster parents because there were so few left, and the Hispaniolan is the nearest relative to the Iguacas, not nearly as threatened with extinction, and kept in the pet trade. The captive program raised significant numbers of Iguacas and after some years conservationists began releasing them to augment the wild flock.

Captive-reared Iguacas have important advantages over their wild-fledged counterparts. They get immediate treatment for diseases and parasites, so almost none of them die from these. They spend their first months in cages that are partly open to the sky, so they learn to deal with downpours from within a safe enclosure.

But once they're released, they have a significant disadvantage. Being raised in a cage means you don't have to learn how to spot and evade predators, and because you also can't understand the language of wild parrots—who are constantly on the lookout for danger—you can't act on their warnings.

The Iguaca's most serious predatory threat comes from the air: Red-tailed Hawks, versatile midsize raptors that hunt across most of North America. You're just as likely to see one plucking a city pigeon off a sidewalk in New York City as sweeping down on a ground squirrel in the Idaho wilderness, but the chance that you'll encounter a Red-tail is greater in El Yunque than anywhere else they've been studied. Ornithologists have recorded more Red-tails per square mile on the island than anywhere else in the species' massive range.

El Yunque's hawks do so well not just because the forest provides abundant nest sites and a large diversity of prey, but because of the local climate. Hunting is exceptionally easy and energy-efficient for them here, thanks to near-constant winds blowing steadily up the slopes of the Luquillo Mountains, creating smooth updrafts for them to glide along as they look down for their next victims. They barely have to flap.

Iguaca conservationists learned that though few captive-bred parrots succumbed to disease and weather after being released in El Yunque, a disproportionate number became meals for Red-tails. In the long term, just as few captive-bred as wild-hatched young survived in the wild. The advantages and disadvantages of being raised in a cage canceled each other out.

These problems did not deter the Iguaca's human caretakers. They soldiered on, spending millions of dollars on research, equipment, and staff, relentlessly solving problems as they cropped up. They set up feeders in the forest to supplement wild birds' nutrition. They attached small transmitters to many of the birds, so they could radio-track them around the clock and locate them if they got into trouble. They cleaned out artificial nests annually and replaced nesting substrates three or four times every breeding season to keep diseases in check. They gave veterinary care when needed. They shot nearby Red-tailed Hawks and Pearly-eyed Thrashers, a bird known to take over parrot nesting sites.

And then, in 1989, they learned the hard way about another damaging dimension of El Yunque's climate. The wild flock had just reached forty-seven birds—its highest population since conservation efforts began—when Hurricane Hugo, a fairly large Category 3 storm, blasted into Puerto Rico and killed half its members. Subsequent analysis has found that the Luquillo Mountains are the most hurricane-vulnerable site in Puerto Rico. The northeastern quadrant of Puerto Rico is three times more likely than anywhere else on the island to sustain a direct hit from a hurricane, and the mountains' shape—like a sharply sloping ramp for hurricane winds to rush up and water to speed down—increases wind damage, rainfall, and the force of flooding.

Hugo's sudden assault didn't keep conservationists down for long. They repaired artificial nests and resumed growing the wild flock bird by hard-won bird. But they began to think that Iguaca had probably not bred in the El Yunque area until colonists leveled all the other Puerto Rican forests and forced them to. The birds probably came to feed here—it's a short flight from the drier lowlands nearby, which would have had good nesting trees until they were logged and turned into city—but a breeding population could simply not survive without skilled, dedicated, and expensive human support.

The last wild Iguaca had become the subjects of an unintentional climate change experiment. They'd been pushed into a hotter, wetter environment than they'd have chosen to breed in, into a place where climate-subsidized predators filled the sky. They needed continuous help from caring people to prevent these sinister green mountains from finishing them off—and then Maria came.

Air, the mixture of gases that makes up the atmosphere, is transparent and odorless, but it is nonetheless still stuff: matter. Because it is matter, it has mass, and because it has mass it is pulled toward the center of Earth by the planet's gravitational force, which keeps it from dissipating into outer space.

The mass of a particular cubic meter of air depends on its temperature. If the amount of thermal energy in that air increases, the molecules in it begin to move and vibrate more, pushing away from each other, making the air less dense and reducing its mass. Warmer air has lower mass and weighs less than cool air, thus the general rule that warm air rises and cool air subsides. Wind happens when warm air rises from the surface and cooler air rushes across from elsewhere on the Earth's surface to replace it; very hot air rises fast, generating strong wind.

One of the most important gases in air is water vapor; that is, water in its gaseous form. A large source of atmospheric water vapor is the surface of the ocean. There, water molecules transition from liquid to gas when they acquire enough thermal energy to vibrate so much that they break the molecular bonds that hold them to other liquid water molecules. Adding thermal energy to water—raising its temperature—makes it evaporate faster.

On September 13, 2017, a large wavelike disturbance formed in the atmosphere above the Atlantic Ocean off the west coast of Africa. Its origin was a patch of ocean a few degrees warmer than normal, from which vast volumes of water began evaporating, transferring huge amounts of thermal energy into the air. This columnar mass of rising air was carried slowly westward, nestled in the trade winds that blow across from Africa toward the Caribbean.

As the mass moved, more water vapor and energy came into it from below, pushing it ever faster and higher into the atmosphere. It began to turn counterclockwise under the influence of the Coriolis effect, a phenomenon caused by the rotation of Earth that makes spinning gyres and vortices of liquid and gas—like the water going down a drain hole—rotate in opposite directions in the Northern and Southern Hemispheres.

The mass became a storm, and as its water vapor rose into cooler reaches of the atmosphere, its molecules lost thermal energy and condensed into water droplets—clouds, which became rain, falling on the outskirts of the storm. Fueled by a constant supply of water vapor and energy from the ocean below it, the storm became increasingly organized, circling around a central point, growing tens of thousands of feet high and tens of miles wide, sucking up and hurling down uncountable thousands of tons of water. On September 17 it developed a well-defined, clear eye and meteorologists gave this new Category 1 hurricane a formal name: Maria.

The next day, as she approached the island of Dominica, Maria passed over even warmer water. Within hours she intensified explosively, rocketing from a Category 1 to a Category 5 hurricane with sustained surface wind speeds of 160 miles per hour, and became the first recorded Category 5 hurricane to make landfall on Dominica. Her short foray over the island deprived her of water vapor and energy, dropping her to a Category 4, but as soon as she moved over the ocean she regained strength, reaching her peak intensity of 175 miles per hour and central air pressure of just 908 millibars.

Thus, as the tenth-strongest hurricane ever recorded in the Atlantic and one of the fastest-intensifying ever, Maria thundered on toward Puerto Rico.

---

The first hours after sunset on September 19 are extremely dark in the El Yunque National Forest. A thick layer of cloud blocks out the night sky. Then the wind begins to pick up, rushing through the forest canopy, bending the trees from the top down, getting stronger all the time. Maria is here. Are the Iguacas holding tighter to their perches, or flying blind to something like safety? Are they calling out to each other? Can they even hear each other over the wind, now reinforced with heavy, stinging, horizontal rain?

By 1:00 A.M. on September 20, just a few hours later, the hurricane's leading edge is tearing full-force into El Yunque. The wind, carrying thousands of tons of flying water, is screaming at well over 120 miles per hour, ripping almost every leaf off every branch and then branches off every tree—and whipping Iguacas into the air, cracking their wings, hurling them into the disintegrating forest or up into black oblivion.

Some Iguacas have somehow made it alive to the forest floor and are hunkered down, soaked through, against the leeward bases of large trees, but by 3:00 A.M. the wind has ratcheted up to an almost inconceivable 150 miles per hour. Now even the oldest trees, durable veterans of hurricanes past, are toppling over and breaking up, punctuating the unearthly thrumming groan of the storm with shocking cracks and thumps. Tons of sharp, rough wood are falling down all around, crushing and smearing Iguaca and all kinds of other wet, huddling creatures into the mud.

It's a pitch-dark hell, and then, later, who knows how much later, when the planet's inexorable rotation forces the Luquillo Mountains toward the sun, it's like there's nothing here anymore. The dawn light illuminates air that is no longer transparent but opaque milky white, that obscures the landscape and becomes thicker and closer as the day advances.

It looks like the thickest of fogs, but it's not fog, it's big water, and it's not still, but hurtling past horizontally. It's as if the air has been replaced by a roaring waterfall rushing sideways through everything. The storm grinds on through the day, with some breaks and a radical change in the wind direction as Maria's trailing side passes over the island. In the forest slopes are becoming water-saturated and collapsing, but the landslides are impossible to see or hear. And then, as the day wears out, the wind and water fade away into the night. Maria is finally leaving.

The morning of September 21 is clear and still, with a blue sky, but the rainforest of two days before is gone. El Yunque is an impassable country of shattered trees—broken, uprooted, and thrown across the hills as far as you can see. The color change is stunning. Two days ago it was green, and now it's mud brown. Almost every leaf has been taken from almost every tree. It's quiet, too, and, with the forest canopy gone, almost unbearably bright and hot. Everywhere stinks. The uncountable tons of vegetation that Maria dumped on the ground, like snowdrifts, are already beginning to rot.

And the wild parrots? Some have survived. After several days, conservationists pick up pings from a parrot's transmitter coming from the town of Aguas Buenas, twenty miles west of El Yunque. The signal moves short distances every few hours: the bird is alive. They get closer and closer, homing in on the survivor, but the woodland that Maria blew her into is also trashed and largely impenetrable, and they can't catch sight of her. They track her for three days, and then the signal disappears. Some Red-tailed Hawks have also survived Maria, and hunting parrots has never been easier for them. The Iguacas' rich green feathers, which once hid them in the leaves, now mark them out brightly against the brown, blasted land.

It takes the Iguaca conservation team a full eighteen days to bulldoze and chain-saw roads and trails open so they can reach the parrot breeding area along the Espíritu Santo river. They set up feeding stations and begin calling up transmitter signals. No birds come to eat. Transmitter pings lead them to twelve birds crushed under wood and another five emaciated corpses—they had starved in the days after the storm. The biologists find no sign of the rest of the flock. After ten days of searching, they give up.

———

The ecosystems of Puerto Rico are not naive to hurricanes, which have likely been passing regularly over the island for millions of years. The island's forests have dense canopies that allow very little light to pass to the forest floor, which as a result is thinly vegetated and open. The forests are dominated by slow-growing but extremely strong hardwood trees; when a normal hurricane hits, their leaves and small branches are blown off, but their trunks and main branches remain. After a storm, the forest floors are covered with a layer of vegetal debris, but this soon rots, providing nutrients for the trees to resprout and recover. Puerto Rico's many types of palm trees are better adapted to hurricanes than most other types of trees. Palms lose their large, old leaves in strong wind, but their strong, flexible, aerodynamic trunks seldom break even in the most severe hurricanes, and they rapidly grow new leaves from their tops after a storm has passed.

One thing Puerto Rico's plants are generally not well evolved to cope with is drought, because regular rain has been a feature of the climate here for a very long time. The island's plants simply don't have the tough leaves and other adaptations that allow them to retain water during dry times because they've never needed them to survive.

But in recent years the island's climate has begun to change, and computer-based climate models predict that these changes will strengthen and become more entrenched. The average air temperature is rising, and rainfall is becoming more erratic, with months of drought interspersed with more intense downpours. Hurricanes are not becoming more frequent, but they are becoming more powerful. Before now, a Category 5 storm might make landfall on the island once a century—but soon, one may come every decade or so.

Puerto Rico has recorded very low rainfall in recent years; as of June 2023, it was in a state of drought. Forest trees have died, especially young trees at lower altitudes. Models predict that by 2100 the lowest-lying forests may become open savannas, seas of drought-tolerant grass dotted with widely spaced trees. Because El Yunque's mountain peaks are high enough to intercept clouds and create orographic rainfall even as the climate changes, many species will survive only at the highest parts of the island. Closed-canopy forest and the species that need it will be compressed into a narrow band around the peaks.

Maria showed dramatically that although most hardwood trees are strong enough to withstand regular hurricanes, they have not evolved to survive major storms. Many large hardwoods in El Yunque were destroyed by Maria, with dramatic ecological effects. With almost all the big trees downed, the upper leafy canopy did not reform. Huge numbers of young trees germinated on the newly sunny forest floor and formed a new-look forest: short, dense, and hot. This has benefited many species of frogs and lizards—whose populations have soared—but has harmed many birds, which need an open understory and the fruit from large, old trees. Palm trees, most of which were not killed by Maria, are coalescing into almost monocultural patches and crowding out other trees in some areas of El Yunque. Even if Iguacas were to recover their numbers, they would struggle to breed in this new forest.

In time, maybe fifty or seventy years from now, enough hardwood trees could become large enough to form a dense upper canopy, shade the forest floor, and thin out the understory to remake a pre-Maria type of forest in El Yunque. But it's likely that another major hurricane will hit the island long before then, and another soon after. Puerto Rico's forests are becoming denser, lower, and hotter.

Intensity matters. A Category 5 hurricane is a completely different thing from a Category 2 storm, ecologically speaking. Because oceans are warming and creating more water vapor, and warmer air can hold more water vapor than cooler air, we can now expect Category 5 hurricanes to rage across the Earth more often than before.

———

At six hundred square miles, Galveston Bay is the largest estuary on the Texas Gulf Coast. It's an expansive, shallow, murky mixing bowl of fresh and salt water: rivers including the Trinity and San Jacinto feed fresh water into its western landward side, and salt water from the Gulf of Mexico passes through gaps in the coastal barrier islands on its seaward side.

But the bay's size can't hide modern humanity's scars. The fifty-two-mile-long Houston Ship Channel has been dredged right across it, a highway along which huge commercial ships can churn between the gulf and the busy terminals of Houston, the muggy oil metropolis that sprawls nearby. The bay's sediments are stained with oil-industry waste; the world's largest concentration of refineries and petrochemical plants was built up here through the oil boom of the 1900s, and dirty urban runoff flows in whenever it rains.

This might seem like a terrible place to be an aquatic animal, but in recent decades rising number of Bottlenose Dolphins have made their home here. Perhaps because of increasingly strong environmental laws and concomitant rising water quality, they've found abundant food. They've also learned to use the artifacts of industry by commuter-surfing container ship bow waves along the ship channel and following shrimping boats to steal food from their nets or pick up discards from the catch; they even teach their youngsters to exploit this source of easy meat.

Galveston Bay's dolphins have complex, fluid social arrangements and distinct personalities. Two males will often form a strong lifelong bond, a brotherhood, sometimes letting females visit to mate, and females will keep their calves close until they mature, but these dolphins don't form strict pods. Individual dolphins will move between groupings or spend time on their own, sometimes gathering in large numbers to socialize or feed and then splitting into smaller aggregations that go their separate ways. Some dolphins keep to certain parts of the bay, while others move around. Of the seven hundred or so dolphins in the bay, about a hundred live more or less permanently near the least-salty topmost part.

Like all animals, dolphins must maintain the concentration of solutes—dissolved chemical compounds—in their body fluids within certain limits, or they will die. You can think of this as the internal "saltiness" of the animal; their blood and other vital body fluids will cease to function properly if they become too watery or too salty. *Osmolality* is the technical term for the measure of how much of one substance has dissolved in another; that is, the concentration of particles in a fluid (like water). Sea water has high osmolality, and fresh water has low osmolality. Ocean saltiness is usually measured in parts per thousand of salt in water.

Salt naturally moves from water of high salt concentration to water of low salt concentration; that is, the saltiness of water evens itself out. The process by which this happens is osmosis, which is passive; it doesn't require the input of energy to occur. Many aquatic organisms need to keep their internal body fluids at a different osmolality than the water around them; they must be either saltier or more dilute than the water around them. To maintain this difference they need barriers structured to allow either water or solutes to move separately across them, and energy to counteract the natural force of osmosis.

Animal bodies have many such barriers, or membranes. Each cell has a microscopically thin membrane around it that controls the flow of solutes and water in and out. Skin can work like this. Kidneys contain membranes that filter waste solutes from the blood, concentrating them in urine and retaining water in the body. Many aquatic animals are built to live in environments that are consistently saltier or fresher than their body fluids, and have evolved physiologies to deal with either too much or too little salt in their environments. Freshwater animals must actively remove water from their bodies but retain salt, and marine animals must get rid of salt while retaining water—but organisms that live in places like Galveston Bay that contain water that can be either fresher or saltier than their body fluids must be able to do both.

Bottlenose Dolphins have extraordinary kidneys. The kidneys of most land mammals, including ours, filter out waste products from blood and return clean water to our circulatory systems. The urine that's created has high solute concentration.

In the salty ocean, Bottlenose Dolphins' kidneys do the same— they excrete concentrated solutes and retain water in the body. They must constantly work to retain as much water as possible: because of osmosis, water is always trying to make its way out of the dolphin's skin into the salty sea.

But when they're swimming in fresh water, Bottlenoses have the opposite problem. Because their body fluids have higher solute concentrations than the surrounding water, water is always trying to come into their bodies through their skin. Their kidneys must reverse the role they play in salt water, producing large amounts of watery urine and keeping salts within the body, and this they do remarkably effectively, having evolved special kidneys that can play both roles—at least for limited periods.

---

Shortly before Maria, there was Hurricane Harvey, which began, just as Maria did, as a disturbance over the eastern Atlantic moving rapidly westward toward the Caribbean. On August 17, 2017, this depression realized sustained winds of forty-five miles per hour and officially became Tropical Storm Harvey, the eighth named Atlantic storm of the season.

Harvey passed across the islands of St. Vincent and Barbados the next day, causing flooding and knocking down trees before continuing west into the Caribbean, where it faltered: wind shear attacked its structure, and relatively cold water failed to supply enough energy to keep it growing. It weakened to tropical wave status in the waters just north of Colombia, then tracked slightly northward across Mexico's large Yucatan Peninsula, which further deprived it of energy and water, degrading it even more.

Harvey's disorganized remnants emerged from Yucatan into the southern Gulf of Mexico early on August 23, heading northwest. In a normal year the storm might have stumbled onward and dumped some rain on the Gulf Coast before finally dissipating, but the sea it now emerged over was unusually warm. Harvey slowed, began feeding on the southern gulf's energy and water, regained its structure—and rapidly swelled into a hurricane.

Gathering strength, the rejuvenated Harvey marched north, making landfall near Rockport, Texas, late on August 25 as a colossal Category 4 hurricane. Then, instead of turning sharply inland as might have been expected, it began lumbering slowly along the coast toward Houston.

For the next three days, Harvey and its spin-off storms cast record-breaking rainfall down on the Houston area. Some places had a year's worth of rain in a day. Some had more than fifty inches in just seventy-two hours. The region's streams and rivers swelled into surging, brown, debris-filled flood, streaming uncontrollably into Galveston Bay, and they didn't stop for over a week, flushing it with three times more fresh water than it usually gets in a year, and eighteen times as much sediment.

---

All cetaceans (dolphins, porpoises, and whales) evolved from an early type of artiodactyl, or even-toed ungulate, a group of hoofed mammals that includes today's pigs, antelopes, giraffes, goats, and cattle. Cetaceans' nearest living relatives are the hippopotamuses, and they arose in the early Eocene epoch when an artiodactyl began evolving an amphibious body and lifestyle, resulting in a strange water-loving creature called *Pakicetus*, fossils of which have been found in fifty-million-year-old sediments in Pakistan.

Most experts consider *Pakicetus* to be the link between early landlubber artiodactyls and the fully aquatic cetaceans. It was about the size and shape of a stout wolf, with a long tail, hooves, and long, strong jaws. Its eyes and nostrils were on the top of its head, so it could view the emerged world while in the water as hippos do, and it may have ambushed prey from the water as crocodiles do. Although it bore little resemblance to modern cetaceans, it had a thickened bone in the skull called an auditory bulla, which meant it could hear underwater as they do.

*Pakicetus*'s descendants radiated into many forms. They became increasingly well suited to the water not by becoming more complex but by discarding genes and simplifying many tissues and parts of the body. They got rid of drag-inducing fur and unnecessary sweat pores by losing the genes that code for hair and sweat glands. They also lost genes for many types of structural proteins in their skin; instead of having different proteins of different strengths and levels of flexibility for different areas and layers, as in human skin, dolphins' skin is made up of many layers of the same type of cells, built from the same small set of proteins.

The outermost layer of human skin—the epidermis—is relatively thin and made of dead cells containing abundant water-resistant fats and keratin. The outermost layer of dolphin skin, however, is alive and permeable to water. Its cells must actively manage their water content—too much water, and they will swell up, explode, and die. Too little water, and they will also die.

Hurricane Harvey turned Galveston Bay completely fresh, not just for a few days, as previous hurricanes had, but for eight whole weeks. Some of the bay's Bottlenose Dolphins swam out to sea, hugging the coastline near the bay, but many remained loyal to their habitual territories, including a male that researchers called #209.

---

Dolphin #209 was a young male, a loner with a territory in the upper bay where the rivers come in. When Harvey turned the whole bay fresh and muddy, many Bottlenoses in the upper bay swam toward the ocean, to the clearer, salty water, but #209, deeply bonded to his area, stayed.

Although many of the upper bay's fish were washed out or swam toward the ocean to stay in their familiar saline milieu, #209 could still find a little food with his extraordinary sonar. But in less than a week his skin began to change. Small pale gray patches, which looked a bit like lichen on rock, began to appear, grow, and spread across his body. He began to weaken and lose weight.

After he'd spent two weeks in fresh water, #209 was covered in the lichenlike lesions. They changed color to purples, yellows, and browns and thickened out, like slimy, dirty outgrowths from his formerly sleek skin. But he remained loyal to his territory, and found enough food to keep going. Whenever he encountered another Bottlenose near him, its skin was also discolored and disfigured to varying degrees.

Dolphin skin cells are sheathed in robust membranes that can control the passage of water and solutes in and out—to some extent. In less-salty water dolphins need an assist from their kidneys to keep their body fluids sufficiently salty.

But in completely fresh water the kidneys must work extremely hard to retain salts inside the dolphin's body—too hard, in fact. After three or four days they start to fall behind, and the dolphin's interior saltiness drops—not enough to kill the animal, but enough to start disrupting its metabolism and its immune system. Less salt inside also means too much water in the skin cells, and they start swelling up, bursting, and dying all across the animal's outer surface.

A compromised immune system plus increasing areas of dead skin cells provide a fertile breeding ground for all manner of microorganisms—fungi, algae, bacteria, many of which were thriving in the now extremely muddy bay. As the invisible creatures breed, they form growing colonies, more and more deeply embedded in the skin. The damaged skin areas become ever less effective at keeping fresh water out and salt in, straining the dolphin's overworked kidneys even more, driving its immune system ever further into disarray.

Freshwater Skin Disease, as this malady has recently been named, wasn't known in dolphins before the late 1990s, when observers started noticing it on some estuarine dolphins and finding horribly discolored dolphin corpses at various places around the world. Scientists first thought it was caused by industrial pollutants or a new type of virus, but they now know that it's triggered by a few days' uninterrupted exposure to fresh water and becomes increasingly serious the longer the exposure lasts.

Most dolphins heal from Freshwater Skin Disease if they are able to access sufficiently salty water before it kills them. In time their skin heals and they appear fully recovered. Four months after Harvey, Galveston Bay was back to its usual salinity, and weeks afterward there was little sign of skin discoloration in its dolphins. Even #209, who had been particularly severely affected because of his insistence on remaining in the freshest part of the bay, appeared well.

But #209 disappeared in 2019, apparently dead at a young age. Scientists fear that if dolphins are weakened by Freshwater Skin Disease, their reproduction and survival may be affected in the long term (this is not yet certain as the relevant studies have yet to be completed).

As droughts grow longer and storms grow stronger, estuarine ecosystems have become much less stable, veering from extremely salty to totally fresh more dramatically than before. Freshwater Skin Disease is now appearing more often in dolphins around the world, from Florida to South Australia to estuaries on the east coast of South America. The emergence of this sickness marks the beginning of a new chapter in the story of climate breakdown and extreme weather, which is also a story of fresh water versus salt.

———

Most Americans think of tallgrass prairie as a Midwestern thing, envisioning the dense, tall grasslands of the north-central United States and adjacent parts of Canada, hot in summer and well below freezing in winter. Few know about a more tropical tallgrass biome, the Gulf coastal prairie, which covered about nine million acres of Texas and Louisiana before European colonists settled there.

The Gulf coastal prairie ran along the sandy coastal plain from Lafayette, Louisiana, southward to just beyond Corpus Christi,

Texas, never extending more than about seventy-five miles inland. Although the region has enough rain and sufficiently deep and fertile soil to support trees, these were prevented from overtaking the prairie by regular fires. Lightning strikes were common, and indigenous people, including the Karankawa Indians, burned the grassland to attract game animals with succulent postfire growth.

The most distinctive animal of this zone is the Attwater's Prairie Chicken, somewhat inaptly named in honor of Henry Philemon Attwater, a British-born naturalist who lived in Texas in the late 1800s and early 1900s. (Feather-splitting taxonomists note that although this bird is roughly the size of a chicken, it's actually a grouse, and conservationists complain that the chicken part of the name makes it sound cheap and domestic, which it isn't.) In 1900 around a million prairie chickens likely inhabited this area, a common icon of the coastal plains.

Prairie chickens are rotund birds with strong, stocky legs. They spend much of their time on the ground, though they fly well when necessary. They're well designed for prairie life, with plumage colored in tight, thin stripes of light and dark brown. A few steps into the tallgrass and they vanish.

Like many other grouse species, male prairie chickens engage in odd rituals to attract mates. In the early spring, small groups of males gather in open areas of the prairie called booming grounds. They face each other and do a ridiculous ritualized dance, which females evaluate before they choose a mate to go off and build a nest with; this gathering is called a lek.

Less than 1 percent of the original Gulf coastal prairie remains today. Being flat, it's easy to plow, and generations of farmers have tilled its fertile soil over and over. It's also easy to build on, and cities like Houston have sprawled out over it, chewing it up and paving it into oblivion. The oil industry and industrial tree plantations have taken most of the rest. Attwater's Prairie Chickens have declined with the prairies. In the 1930s less than ten thousand wild birds remained, dropping to less than one thousand by the 1970s, and less than fifty by 2003.

Since then conservationists have worked to keep the species from extinction, chiefly by breeding hundreds of them every year in zoos. The captive-born young are placed in the patches of grassland that remain, where they're trained to avoid predators and introduced to wild birds to learn the ways of their species. This expensive work kept a population of two to three hundred birds living in the Attwater Prairie Chicken National Wildlife Refuge, a 10,500-acre coastal prairie remnant just east of Houston. But then in 2016 unusually strong spring rainfall flooded every known Attwater's Prairie Chicken nest in the refuge and killed every single egg. With no new young, by the spring of 2017 the refuge's population had dropped to only forty-six.

The remaining males lekked and danced as they always had, about ten pairs were formed, and they nested. Since each nest contains ten eggs, over a hundred young hatched and began to grow through the summer. Then Harvey struck, putting over 90 percent of the refuge at least a foot under water. All the young drowned, and many of the adults retreated to a few small hillocks, low humps in the land just a few inches above the water; sanctuaries from the spreading flood, but also sanctuaries for their predators. Within days bobcats had killed thirty-two of them. After almost two weeks the flood receded and only five females remained. By spring 2018 a mere twelve Attwater's Prairie Chickens were left in the wild.

In 2019 conservationists released more captive-bred young in the refuge. Their efforts initially seemed successful: over two hundred survived their first year out. But then the declines began again, and in 2021 less than a hundred were counted. Heavy rains are becoming more common, and the resulting floods push the

Joshua Tree National Park, Mojave Desert, California, September 2019

Side-blotched
Lizard; Coyote

Gambel's
Quail;
Cactus Wren

Black-throated
Sparrow;
Desert
Cottontail

*All photos* © ADAM WELZ

A "ghost moose" in northern New Hampshire, May 2009 © DAN BERGERON

A mass of Winter Ticks on a Moose
© NORTH DAKOTA GAME & FISH DEPARTMENT

Healthy Pitch Pine stand in Ossipee, New Hampshire, November 2021
© JEFF GARNAS

An adult Southern Pine Beetle
© CAROLINE KANASKIE

Pitch Pine trees killed by Southern Pine Beetle, Long Island, New York, August 2018
© CAROLINE KANASKIE

Adult male Saiga antelope in Chornye Zemli Nature Reserve, Kalmykia, Russia, 2009
© IGOR SHPILENOK

Dead Saiga in the Betpak-Dala, Kazakhstan, May 2015
© SERGEI KHOMENKO

*Top:* Po'ouli (extinct) © PAUL BAKER; 'Apapane © ZACH PEZZILLO; 'Kiwikiu © ZACH PEZZILLO
*Bottom:* 'Akikiki © JACOB DRUCKER; 'Akeke'e © JACOB DRUCKER; 'I'iwi © ZACH PEZZILLO

Bottlenose Dolphins bow-riding in Galveston Bay, Texas, August 2021 © KRISTI FAZIOLI*

Bottlenose Dolphin #209 with Freshwater Skin Disease lesions and growths, Galveston Bay, Texas, April 2019
© SHERAH McDANIEL*

A male Attwater's Prairie Chicken displaying on a lekking ground, Texas coastal plain, June 2017
© JOHN MAGERA/USFWS

Key Deer on Big Pine Key, Florida, September 2019 © ADAM WELZ

*Bottlenose Dolphin images taken under NOAA Fisheries Scientific Research Permit #23203 and provided by Galveston Bay Dolphin Research Program.

Iguacas in captive breeding facility, El Yunque National Forest, Puerto Rico, September 2019
© ADAM WELZ

El Yunque National Forest showing regrowth after Hurricane Maria: dense, almost monocultural stands of Puerto Rican Palm and dense forest interior, September 2019 © ADAM WELZ

Hurricane Maria at its greatest intensity, just before its eye passed over Puerto Rico. 05:45 UTC, September 20, 2017 © NOAA

El Yunque National Forest ten days after Hurricane Maria, September 2017
© TOM WHITE/USFWS

Red Knot feeding on *Loripes* clam, Banc d'Arguin National Park, Mauritania, December 2014
© JAN VAN DE KAM

White Storks wintering in the Kruger National Park, South Africa, March 2009
© ADAM WELZ

refuge's Attwater's Prairie Chicken population down to extremely low numbers from which it cannot naturally recover. A second wild population is now being built up on a private ranch, also on the coastal plain and vulnerable to flooding, and so far that population has grown. But it seems likely that the Attwater's Prairie Chicken will reliably remain in its natural habitat only for as long as young continue being produced in captivity.

---

The Florida Keys are a small group of islands that extend south of the state into the Gulf of Mexico. They're the fossilized remains of sand bars and coral reefs which appeared above the surface of the ocean when the sea level fell about one hundred thousand years ago, during a glacial period.

The fossilized material, limestone, that the islands are made of, is permeable, like a hard sponge. Water can flow through it, albeit fairly slowly. Salty ocean water flows in through the sides of the islands, and when it rains, fresh water moves into the upper surface of the islands. Because fresh water has fewer solutes than salt water, it is lighter, so it floats on top of the salt water inside the hard "sponge." Drill into the ground in the center of a Florida Key, and you will encounter a layer of good, fresh water that is several feet, sometimes many yards deep; this is the island's freshwater lens. Drill deeper, through the base of the lens, and the water will turn salty, fast.

The islands' freshwater lenses sustain a covering of vegetation including many plants that are unique to the Keys. The vegetation, in turn, sustains animals, including an absurdly cute dwarf type of deer called the Key Deer, which is only found here.

The Key Deer evolved from mainland White-tailed Deer when some of the latter were trapped on the Keys by rising sea levels about eight thousand years ago. They had enough food here and a little water—small holes in the top of the islands form little fresh ponds—and they soon began to change to suit this new and rather harsh environment. They became smaller—a lot smaller—as do many large species that find their way to islands.

They're absurdly cute, but people have been hard on the Key Deer. Early white colonial hunters sought it out for trophies, but even though it is no longer hunted, a large percentage of the population is killed by cars every year. It is now classified an endangered species, and a small group of deer protectors, some employed by the National Park Service, looks after the few hundred that remain. This is a temporary arrangement, however. No matter how well conservationists do their jobs, the Key Deer's days are numbered here.

Just ten days before Hurricane Maria tore into Puerto Rico, Hurricane Irma, a category 4 monster, hammered the Florida Keys. Irma generated a six-foot storm surge that sent salt water flowing over the top of the islands. It killed thousands of trees and much of the native scrub and contaminated the natural freshwater sources that Key Deer drink from. The deer were lucky this time; although about 30 percent of their population was directly killed by Irma (hurled into buildings and trees, dragged out to sea and drowned), a few patches of edible vegetation remained alive and local residents set out containers of food and water for the surviving animals.

But even if another hurricane doesn't wipe them out completely, they cannot survive the effects of ocean heating here. As the oceans warm, their water expands; this is one of the chief causes of sea level rise. Salt water is progressively pushing up and into the freshwater lenses of the islands, and they are destined to disappear. Salt will kill much of the vegetation and destroy drinking places. Scientists have predicted that two feet of sea level rise will doom the surface freshwater sources and most of the vegetation in Key Deer habitat; this rise will likely occur between 2060 and 2080, perhaps earlier. Key Deer will have to be moved to the mainland and kept apart from their regular White-tail ancestors to prevent their dwarf genes from being swamped out of existence—or maybe they'll just go extinct.

Sea level rise means that storms now push salt water higher up and further inland than just a few years ago, poisoning wetlands, grasslands, and woodlands along the coast. "Ghost forests" of salt-killed trees are now a feature of the littoral landscape up and down the Eastern seaboard of the United States, thousands of acres of dying habitats advancing inland. Breeding colonies of birds such as the Roseate Spoonbill and Wood Stork are collapsing in the coastal regions of the Florida Everglades as salt increasingly infiltrates that storied ecosystem.

---

Six weeks after Maria, a pair of Iguacas suddenly appeared, circling above the El Yunque captive breeding facility's cages. Two members of the free-flying flock, a male and a female, somehow survived and found their way here, miles from the valley of the Espiritu Santu. The elated staff immediately laid out food for them, and the hungry birds ate well, roosting in trees near the cages that night. Over the next month they established a territory around the facility and moved into a new artificial nest that was quickly put up for them. They stayed close to one another, and were looking as if they would breed until the female was apparently taken by a Red-tail.

Breeding center staff lured the surviving male into a live trap and identified him as a cage-born bird less than two years old. They cared for him in the facility until January 2020, when they released him along with nineteen other Iguacas to form a new free-flying flock based around the breeding cages, not in the old territory. (When a big hurricane comes again, as it will, conservationists want to be able to immediately feed any birds that may live through it. No one wants to lose another flock or more decades of work.)

The last male of the old flock may well be able to teach the new releases something about forest living, like which wild fruits are best and how to look for danger. He's already found a mate and raised young. But he won't be able to pass on the old Iguaca language or the parts of Iguaca culture that lived along with it. He'd not spent enough time with the true wild birds to learn it.

We humans know that the old language existed, and we can pass recordings of it to our descendants. But even if we could teach future generations of Iguacas to make the sounds encoded on our magnetic tapes, flash cards, and hard drives, we could not resurrect the parrot language that began evolving over half a million years ago and was passed from generation to generation without fail. Our recordings are little more than echoes of sentences without speakers or dictionaries to explain them. Not one of us, nor the living Iguacas, knows what they mean, and in September 2017 Maria made sure that we never will.

# 4

# Morphing Migrations

T he coast of Mauritania runs along the western edge of North
Africa. Here the vast Sahara ends abruptly at the Atlantic
Ocean, desert sands giving way to open water in a matter of yards.
This shoreline can appear almost dead from a distance—the sea
surface is flat and featureless, the landward side sunblasted, arid,
with little vegetation. But over 150 miles of its length is included
in the Banc d'Arguin National Park, and as the tide drops on this
sunny afternoon in September 2002 you can begin to see why it's
considered a Mauritanian and international treasure; hundreds of
square miles of rich, gleaming mudflats and seagrass meadows
emerge into the hot, rippling air, out of which a tight, fast-flying
group of fifteen small gray birds appears.

The birds wheel around together and flutter down, extending their long legs to land on the mud. These young Red Knots have just concluded an epic trek which began in northern Siberia and passed through Europe. Over the course of their journey, they've consumed over an ounce of their five-and-a-half-ounce body mass to power their ever-pumping flight muscles. And now, having lost so much weight, they need to eat. They quickly tidy their feathers and begin probing their long, thin shorebird bills into the wet surface.

A few evenings ago these birds were part of a thousand-strong flock of Red Knots feeding on the tidal flats of the Wadden Sea on the northern Dutch coast. Some of the flock had been calling to each other and making short test flights together, seemingly negotiating good partners for the final legs of their passage to the Banc d'Arguin. At sunset, groups of ten to twenty birds separated out and bunched up together, flying away into the darkening sky.

They navigated south over France, Spain, and Morocco, driven by a relentless urge to move along a route established by their ancestors. They battled headwinds and sidewinds and, near Gibraltar, lost a few of their kind to waiting, circling missile-like falcons. At night they oriented themselves by the stars. By day they used a sun compass calibrated with an uncannily accurate sense of time. They navigated by instinct, their course encoded in their genes; as young juveniles flying alone on their first long journey, they had no experienced adults or memory to guide them.

Flying from Siberia was hard on the birds, but they face another challenge in Mauritania: they must remake their bodies for this different world. To thrive in the Banc d'Arguin they must feed on the *Loripes* clam, the size of an American dime coin and the most abundant, protein-rich knot food here. The Red Knot has evolved a unique tool to find these important clams, which live by the millions just under the surface of the mudflats.

Every Red Knot's long, thin bill is tipped with clusters of minute pressure-sensing organs called Herbst corpuscles. When a knot probes into wet mud, its bill displaces a small amount of water, pushing it away into the surrounding sediment. If there are no hard objects nearby in the mud, the water moves away relatively unimpeded. But if there is a hard object like a clam nearby, the water's movement is obstructed, and this increases pressure on the knot's bill tip. The knot can thus sense a nearby clam's presence and efficiently find this invisible prey.

Finding little clams is one thing, but getting to the soft flesh inside their hard shells if you have a fine, weak bill like a Red Knot's is quite another. During their first short months of life in Europe, the young birds subsisted on small insects and other soft-bodied invertebrates that presented no challenge to them. Now they must restructure their bodies, which they have evolved to do automatically on arrival in Mauritania; they begin to break down their pectoral flight muscles, which had been enlarged for their long migration, and reallocate those resources to growing a large gizzard, a stomachlike chamber in their digestive system that's lined with muscles strong enough to crush mollusks.

Juvenile knots stay on the Banc D'Arguin for about twenty months as they mature. They rely on other, less nutritious foods to sustain them while their bodies change and they become more practiced at finding and digesting clams. By the time they reach adulthood, *Loripes* should make up the majority of their diet, and they should be thriving, strong, and ready to fly north to breed for the first time, just as thousands of generations of their ancestors did.

But these birds are not thriving. Only a hundred thousand were counted here in 2022 compared to the half million or more who regularly wintered here in the 1980s. *Loripes* clams still populate the tidal flats, but nowadays the knots feed heavily on low-protein Dwarf Eelgrass, a plant that grows in the mud. This isn't a matter of choice for them: they can't easily reach the clams anymore because in northern Siberia, well over five thousand miles away, the climate is changing fast.

---

Many animals move to different areas during their lives, and these movements can take different forms. The young of some species move away from where they were born to establish their own territories, a type of one-way movement called dispersal. Other species exhibit nomadism—they move around unpredictably, usually to use unpredictable resources—like desert animals that travel toward distant, irregular rainstorms in the hope of finding new grass. And then there is migration, the predictable, recurring seasonal movement of animals between different places, which is normally undertaken by all members of a species or population (it's not an individual thing).

Migration has dangers. Leaving the familiar environs of home and trekking through unknown territory can get you killed. But migratory behavior arose long ago—fossils show that animals have migrated for at least tens of millions of years—and it's endured because animals gain advantages, including access to bountiful seasonal food sources, that outweigh the disadvantages of regularly moving to different places.

Birds are the ultimate migrants. About one-fifth of all bird species migrate, and they can move faster and farther than migrant mammals, insects, fish, or any other group of animals.

Fossil records and genetic evidence show that being sedentary was the original mode of life, so it's easy to envision how migratory behavior could evolve: Imagine a nectar-feeding bird species that lives along low river valleys year-round, subsisting on perennial flowers. Then imagine that the climate slowly begins to warm—because a glacial period is ending, perhaps. The snow on nearby mountainsides begins to melt away in summer, creating new habitats in which fast-spreading annual flowers can bloom for a few months every year. Some members of the bird species see these new fields of summer flowers and fly a mile or two upslope to investigate, and discover that the new plants produce more nectar than their traditional valley-bottom flowers. So they set up territories and breed in the new meadows, producing more and healthier young than the birds who remained in the valley bottom. When winter comes, dumping snow on the slopes and killing the annual flowers, the birds simply move back to their old low-altitude haunts for the season.

As the climate continues to warm over the next few years, the new seasonal flowers grow farther upslope and farther away from the equator. In summer the birds move to where the best blooms are, but are always forced back to their original range when winter bites. Because the birds that migrate have more and stronger offspring than their sedentary species-mates, migratory behavior and its associated genes become increasingly more common in the species. In time the sedentary lifestyle might entirely cease to exist in the species and it will become completely migratory, moving farther and farther each year between its expanding summer breeding range and its old—now winter—range as the environment changes. "Farther and farther" can eventually become thousands of miles, as evidenced by the many long-distance migrants that fly across the planet today.

Migrants need special tools over and above those required for sedentary life, because they need to be able to know where they are geographically, even if they are in an unfamiliar place, and they need to be able to reliably follow the same, often very long, routes, year after year. Many types of animals have spent millions of years evolving extremely sophisticated tools for migration, some of which scientists have come to understand, while others remain at least partly mysterious.

Studies have demonstrated that birds can locate themselves and navigate using the stars, the position of the sun in the sky, their sophisticated smell memory, memorized visual maps of the landscape, and their ability to tell time extremely accurately and to sense differences in Earth's magnetic field. Most species appear to integrate various tools and senses as they migrate.

Many large birds with large wings migrate during the day, because they can soar on masses of rising air and winds generated by the sun-warmed ground, which reduces the energy requirements of travel. Many small birds migrate at night; their wings aren't large enough to soar on, and they must flap continuously, a task made more difficult by unstable daytime air.

Experiments with night-migrating birds in planetariums show that they continuously observe star movements to locate the celestial poles. The stars above the North and South Poles barely move as Earth rotates, but the stars above the equator speed across the sky (relatively speaking). Birds find north and south by identifying the slowest-moving patch of stars above them, and adjust their flight bearings accordingly.

Day-migrating birds use the azimuth of the sun, combined with an extremely accurate internal clock, to find their bearings, much as early seafarers used their sextants. Many birds also have very good visual memories of the landscape, and use visual landmarks to repeat migration routes and find their destinations; individuals will find their way back to the same tree, the same fence post, the same nest box, year after year, from thousands of miles away.

Perhaps the most remarkable ability that migrant birds possess is assessing latitudinal differences in Earth's magnetic field. Earth's molten core generates a magnetic field around the globe, which creates lines of force that run between the magnetic North and South Poles. These lines are parallel to the planet's surface at the equator, but angle more sharply downward as they approach the poles. (You can imagine these lines erupting into the atmosphere from the North Pole and curving southward, maintaining a constant altitude over the equator, and then plunging sharply toward the surface as they approach the South Pole.) Earth's magnetic field is also stronger at the magnetic poles than at the equator.

It's long been known that birds can use Earth's magnetic field to orient themselves. Decades ago scientists placed migrating birds in dark rooms, without any sight of the sun or night sky, and they flew in the correct direction. It's further been assumed that birds' magnetoreceptors are in their eyes, because birds with damaged eyes or damage to the part of the brain that processes visual information can't sense the magnetic field. Other experiments showed that live birds that had red light shone in their eyes were unable to sense magnetic fields—different-colored light appears to activate or deactivate magnetoreceptors—and birds kept in dark rooms saturated with "electrosmog" (low-level magnetic fields from electronic equipment) were also unable to.

But it's only very recently a possible explanatory mechanism has been discovered via lab experiments that showed that weak magnetic fields like the Earth's can affect chemical reactions in the retinas of birds' eyes. Birds might, in some strange way, be able to "see" Earth's magnetic field to determine how far they are from the poles.

The mechanism is complex and not yet fully understood, but research published in 2021 describes how scientists produced a quantity of cryptochrome, a type of protein found in night-migrating birds' retinas, then exposed it to blue light and placed samples in magnetic fields of different strengths. Different chemical reactions occurred in part of the cryptochrome molecule, depending on the strength of the magnetic fields. The scientists found that different chemical reactions occur in cryptochrome depending on the strength of the magnetic field it's placed in. Cryptochrome can thus theoretically send messages into birds' visual sensory system that would allow them to assess the strength and possibly the angle of inclination of magnetic fields, which would help to figure out their position on Earth.

———

Ornithologists have identified and named six different subspecies of Red Knot, all of which breed in the Arctic during the northern summer and migrate to temperate or tropical regions for the nonbreeding season. Genetic evidence indicates that they evolved from a single ancestral type which split into two lineages about thirty-four thousand years ago. After the Last Glacial Maximum (twenty thousand years ago), as the Earth warmed, the ice retreated, and more habitat opened up, birds from these two populations evolved new migration routes, new breeding patterns, and subtly different physical attributes, thus splitting into the six subspecies of today. Each subspecies has a well-known breeding and nonbreeding range and travels well-established routes between them.

The Afro-Siberian Red Knot (*Calidris canutus canutus*), the subspecies at the center of this story, migrates annually to the Taimyr Peninsula in far northern Siberia to breed. For thousands of years its members timed their northward flights to arrive at this massive hammerhead of low-lying land on the edge of the Arctic Ocean just as the winter snow begins to melt and spring begins to tease the tundra vegetation—and the vast numbers of small invertebrates that live in it—into life again.

As with many migrant species that breed in strongly seasonal climates, Afro-Siberian Red Knot reproduction is a speedy, tightly scheduled affair. As soon as the birds arrive at their breeding grounds, the males occupy territories and start attracting females, and within a week the birds have paired up, made nests—just small scrapes on the ground—and begun laying a clutch of (usually) four eggs at the rate of one egg per day. Males and females alternate egg-incubation duties, and their young hatch three weeks after the clutch is complete, just when tundra insect populations have, for many years, reached their spring peak.

As soon as the young emerge from their eggs, mother knots leave them and begin flying south. Males are left alone to raise the babies, which start out as tiny balls of fluff that are perfectly camouflaged in the low tundra vegetation. Although the hatchlings are precocial—able to walk around and find their own food almost as soon as they break out of the eggs—their tiny size means they have little thermal inertia. They lose energy very quickly to the surrounding cold air, and they're unable to generate enough metabolic thermal energy inside their bodies to make up for the losses. So for their first ten days of life their father must periodically gather them together under his wings and warm them up. Their father also keeps a lookout and warns them of approaching predators, like Arctic Foxes, uttering a sharp peep that sends them down into the tundra, motionless and nearly invisible, until danger passes.

The young feed voraciously on tundra insects. They reach full size and are able to fly in just three weeks, at which point their fathers fly south, leaving the young alone to bulk up on the last remaining invertebrates of the short Arctic summer. A couple of weeks later, with the snow and ice closing in, the young birds too start flying south, their genetically coded instincts routing them via the vast Gdansk wetlands on the Polish coast and the Dutch Wadden Sea to Mauritania.

This schedule worked well for Afro-Siberian Red Knots for a very long time. But in recent decades the Arctic has been warming three to four times faster than the average for the rest of the globe. Spring snowmelt has been arriving progressively earlier on the Taimyr Peninsula; scientists found that between 1983 and

2015, it advanced by half a day per year on average.[¶] Spring warming also progresses erratically, sometimes interrupted by intense cold snaps which kill many insects. So the spring insect peak not only comes weeks before it used to but also produces insects far less reliably than before. Recent research has found that crane flies, insects that look a bit like very large mosquitoes, are particularly important for Red Knot nutrition. Crane flies are also very sensitive to changes in climate; their populations have been peaking far in advance of Red Knot hatching.

Fewer insects means that male Red Knots have less energy to keep their young warm and safe in their vital early weeks, and many fathers abandon their young prematurely to move south to more stable feeding grounds. Young knots are more likely to be malnourished, and many perish before they learn to fly. Ornithologists found that juvenile knots that arrived in Mauritania in 2015 were a remarkable 20 percent smaller and lighter than those in the early 1980s. Their average beak length also shrank, though by only 10 percent; they were about three millimeters, about one-eighth of an inch, shorter than before. Afro-Siberian Red Knot beaks range in length from a little under thirty to a little over forty millimeters, so a three millimeter drop in the average length means that more knots now have beaks at the shorter end of the range.

Bill length theoretically means little on the tundra—shorter-beaked knots should catch insects just as well as longer-beaked knots—but it becomes critical once they arrive in Mauritania. Although their special pressure-sensing organ still works, knots with shorter bills can't probe quite deep enough into the mud to reach the majority of *Loripes* clams; ornithologists have found that knots with thirty-millimeter-long beaks can only reach about a third of the *Loripes* clams in the mudflats, whereas birds with

---

[¶] This advance has continued since 2015, but estimates of its recent rate are not yet available.

forty-millimeter-long beaks can reach more than two thirds. Instead of shifting to this high-protein food that their species is uniquely adapted to find, an increasing number of young Red Knots is forced to continue eating a lot of Dwarf Eelgrass and smaller, less abundant species of shellfish during their time on the Banc D'Arguin. They simply don't get enough nutrition, especially protein, and over months they start wasting away. Many die.

Although the Afro-Siberian and Red Knot average size is going down, its members are under sustained evolutionary pressure to keep their bills as long as possible. Birds that can still reach a lot of *Loripes* clams survive and breed far better, which is why bill length is declining more slowly than body weight. As one researcher wryly remarked, we are entering the age of the Pinocchio knot.

A family of White Storks—mother, father, and two newly fledged youngsters—is earnestly stalking dinner in a muddy ditch on the edge of the tiny village of Böhringen in southern Germany. They're blithely unaware of their status in European human culture as symbols of birth, the large black-and-white bringers of babies, although they have long lived among people. For over a thousand years storks have nested on the rooftops of European villages; a particularly large and long-used nest graces the steeple of the old church in central Böhringen. Every few minutes one of the birds jerks its spear-like red bill into the water, pulls up, and then casually tosses a frog into the air before catching it and swallowing it headfirst, the frog forming a small, wriggling bulge that slowly pulses its way down the stork's neck and into its belly.

The stork family is fattening up because temperatures are dropping. The big birds must leave before the ponds and marshes that produce their fish, amphibian, and insect prey freeze, or they'll starve. They're pushed southward by winter, like many other bird species that lose access to their food as the cold intensifies. They're preparing to fly south along an ancestral migration route that traverses Spain to the Straits of Gibraltar, where they'll cross the Mediterranean to Morocco, overfly the Sahara, and spend the nonbreeding season in the Sahel, the strip of West Africa savanna sandwiched between the desert and the coastal rainforests of that region. Both stork parents have endured this challenging trans-hemispheric migration for twenty continuous years now, returning each spring to Böhringen's old church steeple and their massive nest, itself inherited from previous generations of storks who built the dense pile of sticks up over centuries.

Thousands of White Storks fly from Europe to Africa each year, often grouping into massive flocks that move together. They can go as far south as the southernmost tip of the continent, feeding on small reptiles, mammals, and insects in the hot drylands, grass-lands, and savannas that cover so much of the terrain. It's not unusual to see large groups of White Storks marching among herds of zebra and wildebeest, plucking up creatures disturbed by the grazing mammals.

Although there is a genetic component to stork migration—youngsters traveling alone on their first migration will instinctively fly to Africa—there is a strong learned component, too. After their first migration, storks can repeat their first journey with incredible accuracy for the rest of their lives, obviously having memorized it carefully. Although juvenile storks can and sometimes do fly south alone, there are real advantages to migrating with their parents and adults, who can show them how to find giant thermals and soar within them, avoid dangerous powerlines and hunters, and stop in the right places to feed. A single small mistake means death; most juvenile storks don't make it through their first migration, though their odds of survival improve when flying with older birds.

But this year, 2019, the Böhringen stork family will have a different journey south. While thermaling over central Spain on an unseasonably warm autumn afternoon, they'll see the black-and-white feathers and bright red bills of hundreds of other storks feeding far below. They'll spiral down, landing with a hop and a bump in the middle of an overflowing landfill site littered with chicken heads from a nearby slaughterhouse and bursting bags of expired pork sausage from the local meat factory, and be inducted into an orgy of stork feasting. The Spanish winter won't get as cold as it used to, so the parents will have no strong incentive to go farther south, though after some time the youngsters' instincts will overcome them and they'll travel on to the Sahel—though only for their first migration. In subsequent years they'll stay with the older storks in Spain, and never see Africa again. The birds will remain at the dump until spring, when they'll head back to their village in Germany on grunge-smeared wings.

This migration pattern has become increasingly common since the year 2000. Many White Storks in western Europe are becoming short-distance, intra-European migrants. Their culture is changing thanks to warming winters and a newly discovered, reliable source of food.

But White Stork culture isn't changing only in the Northern Hemisphere. In recent decades a small number of the birds have begun staying in the very southernmost part of Africa, near Cape Town, instead of migrating back to Europe. A few pairs have bred here and their young stay here all year round too. They feed in small farm ponds, just like the European birds, and on garbage dumps, just like the European birds.

Migration comes with a cost. Every mile traveled requires energy and adds risk, and when movement is no longer a matter of life and death—as with a freezing winter that prevents access to food—that cost may no longer be worth paying. Individuals override their migratory impulses and habits and stay put, and if they breed successfully, they can evolve to new, sedentary species.

This may now happen with the White Storks breeding in South Africa. They may pass on their sedentary lifestyles to their offspring, and because the breeding grounds of migratory and sedentary populations are far apart, genes will quit flowing between them. Also, because the populations are under different evolutionary pressures, they will evolve to become distinct species. It's already happening to North American Barn Swallows: since the 1980s, some migrants have stayed on their wintering grounds in Argentina and begun breeding there. These Barn Swallows head north when the southern winter arrives, but they don't cross the equator; they simply migrate to tropical South America for the cold season before heading back to Argentina again. Although they still look identical to their recent ancestors that breed in North America, DNA studies show that their genes are already changing—they're in the early stages of becoming a new species.

The Black Harrier, one of Africa's most attractive raptor species, arose this way. Its dramatic black-and-white plumage looks quite different from the overall gray of its relative, the Pallid Harrier, but studies show that they're genetically very similar. The Black Harrier is a sedentary species, confined to southwestern Africa, whereas the Pallid Harrier breeds across central Eurasia and migrates to sub-Saharan Africa and South and East Asia. Genetic evidence indicates that about a million years ago, some Pallid Harriers stayed in southwestern Africa and became today's Black Harrier.

There are many examples of such migrant/sedentary species pairs around the world. In fact, genetic and taxonomic research shows that migratory species have split up and given rise to daughter species far more often than nonmigratory species have done.

It's a safer bet to repeat a migration route that's proved successful in the past than adventuring along a randomly selected new route, and many species have strong genetic and behavioral mechanisms to conserve the memory of old routes through generations.

The young of many larger bird species—like storks and eagles—learn a lot about their migration routes from adults; most of these species migrate in mixed-age flocks. But in many small bird species the urge to migrate and general migration directions are more strongly encoded in their genes, and juvenile birds often undertake their first migrations alone, before their parents leave the breeding grounds.

Blackcap Warblers are common tiny songbirds that breed in Europe. They're gray overall, and the males have black tops to their heads, females and young birds sporting red-brown caps. When their insect food begins its temperature-related decline in late autumn, they fly south; the Blackcaps breeding in western Europe pass across the Straits of Gibraltar, and those breeding in eastern Europe fly along the eastern Mediterranean coast, through Lebanon and Israel, to reach Africa, and then disperse across the continent. A few months later they backtrack along the same routes to get back to the areas where they were born so as to breed in the European spring.

Scientists have raised Blackcaps in captivity and have shown that young birds from eastern and western populations have strong genetic impulses to fly in the correct direction for the correct number of days to get them to either the Straits of Gibraltar or the eastern Mediterranean coast without adult guidance. Scientists have also cross-bred Blackcaps from different populations, taking one parent from a population that goes westward around the Mediterranean and another parent from a population that takes the eastern route. Their hybrid offspring turn out to be genetically programmed to fly a route halfway between their parents' routes—straight across the middle of the Mediterranean. Any Blackcap attempting such a long water crossing will likely drown, its genes removed from the gene pool; there is strong evolutionary pressure to remove genes that encourage dangerous travel.

But since many migratory species have survived through millions of years of major environmental change by changing their migration patterns, it stands to reason that it's possible. Changing migration routes isn't a rational, preplanned thing—it comes about through accidents.

Each generation of a migrant species typically contains a few individuals who don't follow the script, who travel new or atypical routes, perhaps because of genetic code corruption or by being blown off course by severe weather. These are vagrants, and nearly all vagrants are juvenile birds migrating for the first time. Birdwatchers love vagrants, because it's exciting to find a species in your own home patch that's usually found far away, but their lives often end quickly and tragically; they starve, freeze, or overheat, or are killed by unfamiliar predators. Very occasionally, however, a vagrant wanders into a new wintering area that suits it well, and manages to fly back to its species' traditional breeding ground, perhaps returning again to its new wintering area for the next nonbreeding season with its offspring in tow. If the new destination sustains the species at least as well as the traditional wintering areas, the number of individuals flying there may increase over the years, and a new migration route will become established.

In the 1940s British birdwatchers noticed a tiny handful of Blackcaps arriving in England at the onset of winter and spending the cold months there, which was very strange because Blackcaps traditionally left Britain at that time of year. Although the species usually subsists on insects, these new wintering birds lived off suet and seeds in birdfeeders put out in English backyards. Over the following decades their numbers very slowly increased, and today it's common to find overwintering Blackcaps all over the British Isles. Researchers have tracked them and discovered that they breed in central Europe—in Germany and nearby countries. Somehow small numbers of central European Blackcaps started migrating west instead of south, and the combination of reliable human-supplied food and steadily increasing winter temperatures supported these "lost" birds with the bizarre east-west migration route and allowed their numbers to grow. Most Blackcaps still fly to Africa, but as climate breakdown begins to bite, that traditional route may become less suitable because of increasing drought, which hammers African insect populations, and more extreme storms, which kill migrating birds.

The Blackcap Warbler isn't the only species developing a new and strange migration route. Richard's Pipit is a drab olive-beige songbird that breeds in the short grasslands and tundra of north-eastern Russia, around Siberia. Come the northern winter, it has traditionally flown south to the Indian subcontinent and Southeast Asia. In the 1980s and 1990s French birdwatchers began to find a tiny handful of birds— in most years fewer than five individuals— overwintering in their country. They were nearly all juveniles and assumed to be regular "doomed" vagrants, but then their numbers steadily began to rise. Since 2015, over a hundred Richard's Pipits have been spending the winter in one area of southern France. Scientists placed tiny tracking devices on some of these birds and found that they fly all the way back to Siberia to breed, afterward returning to precisely the same small French fields where they spent the previous winter, a pattern they repeat year after year.

Even though the distance between Siberia and France is thousands of miles longer than the distance between Siberia and southern Asia, the new migration route seems to be working for the species. Droughts, habitat destruction, and pesticide use are increasing in India as winters in southern Europe become milder.

It's likely that an increasingly large percentage of the Richard's Pipit population will fly the long east-west route in the future.

———————

Timing is a vital element of migration. Leave too early or too late, and you might run into deadly weather or fail to find food at your destination. Hundreds of millions of birds migrate between Europe and Africa each year, their survival dependent not only on following the correct routes but also on correctly timing each stage of each journey. Migratory birds have extremely accurate internal clocks to help with this; many species have been shown to be able to "tell the time" with an accuracy of about five minutes. They can know not only what "time of day" it is—obviously birds don't express this in terms of hours and minutes like us—but also how long a day is.

In the mid-1990s, scientists kept migrant birds in sealed rooms lit only by artificial lights and created a fake autumn by gradually shortening the time that the lights were turned on every day. As soon as the fake day length matched the natural day length at the time of year that the birds normally left their breeding grounds, the birds inside the rooms began to hop around more than normal and try to fly away. German naturalists of the 1700s coined the term *Zugunruhe*, which translates to "migratory anxiety" or "migratory restlessness" for this behavior after observing it in wild birds. It persists for days, and stimulates the birds to get moving. The timing of the onset of *Zugunruhe* appears to be genetically determined in many species, just as basic route instructions are.

Because birds have no way of knowing the weather or habitat conditions that await them hundreds or thousands of miles away, they cannot adjust their initial departure time according to conditions at their ultimate destination. Every long flight—which often passes over hostile territory—is thus a flight of faith. If conditions remain the same year after year, it makes sense to repeat previously successful migration timing, just as it makes sense to repeat previous routes. But when climate changes, it can pay to leave earlier or later than your population or species traditionally has.

Individuals of species that breed in the far Northern Hemisphere benefit by arriving at the breeding grounds as early in the spring as possible. The early birds get to select the best nesting territories and have the most time to court mates—although arriving too early, before the breeding grounds are warm enough, can be dangerous. As Earth warmed during interglacial periods in the Pleistocene era, snowmelt and spring would come progressively earlier to the Arctic, and individuals and species that bred there would gain advantages if they arrived earlier, too. Fossil evidence shows that many migrants thrived despite the waxing and waning of the ice sheets during the Pleistocene, so they were clearly able to adjust their migration timing as conditions changed.

Birds with genetically encoded migration patterns will slow their journeys slightly if they encounter extremely good or bad conditions along the way. Willow Warblers—which are even tinier than Blackcaps—and many other small insectivorous song-birds will sometimes pause their northward flights across Africa if they encounter good rains, in order to take advantage of rain-boosted insect populations. But they can delay only so much; their schedule has limited flexibility, and after a few days of bulking up they will continue north. Similarly, if they encounter dangerous weather, such as a big storm, they may hunker down for a couple of days instead. But these on-the-fly adjustments don't usually change individual birds' departure timing during their subsequent migrations.

Population-level changes to migration timing usually happen like route changes do, via individuals who accidentally break the rules because of genetic mutations or for other reasons. When conditions are changing, misfits can gain advantages to be passed on to their offspring.

Red Knots have been arriving on their Taimyr Peninsula breeding grounds earlier than in the 1980s; they are evolving to advance their arrival time as the climate warms, but not fast enough. Their rate of arrival advance is only half of the rate of snowmelt advance, an average of a quarter of a day per year versus half a day per year. They've tried to compensate for this by nesting at progressively higher altitudes than before, selecting cooler sites where insect emergences are slightly delayed; each ten meters of altitude delays insect emergence by about a day. The Taimyr's tallest hills top out at only fifty meters above sea level, and the knots have already, in recent years, started nesting there. They have nowhere higher or cooler to go.

Why haven't the knots kept up with the ever-earlier snowmelt? Perhaps it's because of the extremely high cost of arriving in the Arctic just a little too early. The Taimyr Peninsula produces insect food for Red Knots after snowmelt, but there is absolutely nothing for the birds to eat when snow and ice cover the ground. If Red Knots arrive at their traditional breeding grounds just a few days too early, they will not merely suffer inconvenience or reduced breeding performance; they will starve or freeze to death.

The Red Knot as a species had to have been somewhat schedule-flexible to survive past climate change, but knots with just a little too much flexibility in their migration schedules would have rapidly dropped out of the gene pool. It stands to reason that the Red Knot population could over time have become dominated by individuals carrying genes that constrained the extent to which they could deviate from established migration schedules; this would function as a governor on the rate at which the population's schedule could change.

Juvenile knots don't always follow their parents' schedules; they might migrate a little earlier or later than their parents. But research suggests that individual Red Knots keep to the same migration schedule throughout their lives; year after year, they stick to the same departure dates that they flew by in their first migration. So Red Knot migration schedules only change between generations, not within generations.

Current human-caused Arctic warming is proceeding much faster than any natural warming episode that Red Knots have experienced in their species' entire existence. It's likely that the mechanisms that successfully moderated their migratory evolution in the past are now preventing them from changing fast enough to survive.

───

Climate breakdown is affecting vast numbers of migrant animals all over the world. Dozens of North American bird species are arriving earlier at their breeding grounds. Many species of long-distance migrant birds are evolving smaller bodies and longer wings, to move farther, faster, on less food. Marine fish are moving closer to the poles. The famous wildebeest migrations in east Africa are breaking down as rainfall becomes erratic and droughts deepen.

Fossil evidence indicates that during past, natural episodes of climate change, migratory bird species were more likely to survive than sedentary species, likely because their predisposition to relocate to more suitable habitats. But current conservation assessments show that migrants are more likely to be threatened by extinction than sedentary species; the rules are being reversed. Not only is climate change now proceeding more rapidly than in the past, but many other human-caused threats are simultaneously affecting birds: pollution, overhunting, habitat destruction, and so on. Migrants rely on many different places and ecosystems during their lives, and it only takes the destruction of one of these stepping stones to threaten their survival.

Environmental changes affect not just the migrant species themselves but also the ecosystems with which they interact for part of each year. Take for example the leaf warblers, which are tiny, drab gray-green birds of the genus *Phylloscopus*. (They're the bane of birdwatchers because the various species look so similar that they can be impossible to tell apart in the field.) Eighteen species of these diminutive migrants spend the nonbreeding season on the Indian subcontinent, south of the Himalayas—they breed across Eurasia, north of that storied range.

A biologist who researched leaf warblers in an Indian forest found that for 200 to 250 days of each year each hectare of forest hosted six to eight individual warblers. They fed on insects, mostly leaf-eating caterpillars. He observed that each warbler ate an average of three insects per minute during waking hours, or about 1,980 insects per day. Six warblers would thus remove almost 12,000 insects from each hectare of forest every day they were there. Billions of *Phylloscopus* warblers likely winter on the Indian subcontinent every year; what will happen to the trees if they no longer fly here? Will the caterpillars run rampant, defoliate, and ultimately cause the collapse of forest ecosystems? It's impossible to know, of course—maybe, just maybe, sedentary birds would increase and take their place in the food web—but it bears serious consideration.

Migrants link distant species and ecosystems together. They have important relationships with other species in different places, species that would otherwise have nothing to do with each other. They are actors in many different worlds.

Migrants also urge us to recognize that species relate to each other within ecosystems in terms not just of space (where they physically are in relation to each other, and how much space they occupy) or their positions in a food web (consumer or producer, predator or prey) *but also in terms of time.*

Relationships between migrants and other species can fall apart if their actions fail to coordinate in time, which can ultimately lead to extinctions and ecosystem collapse: birds miss insect emergences, and so their young are malnourished. Large grazing mammals can't run short of grass because the rains fail to fall on schedule, so they don't breed, and the predators and scavengers who live off them fade away too. Leaves emerge too early in the spring, before the caterpillars who rely on them for food have hatched, and so the butterflies that pollinate many species of wildflowers disappear, and so on. Sometimes species can adjust their schedules to resynchronize relationships, but often they can't, especially now, when the globe is changing so fast.

―――――

A few decades ago, when the Afro-Siberian Red Knot population was stable and its Arctic breeding area wasn't yet noticeably warming, the subspecies would normally lose about 15 percent of its adult population every year to predation, migration accidents, and natural death. Thus, to maintain its numbers, juveniles equivalent to at least 15 percent of the population would have to survive to adulthood every year. Observers would would expect that at least 15 percent of a healthy population would be juveniles.

Scientists censused the knots wintering on the Banc d'Arguin in early 2022, as they have done annually since the early 1980s. Besides confirming an ongoing decline in numbers, they found that less than 1 percent of the birds were juveniles. They also observed a rapidly growing sex imbalance: in 2022 there were almost three females for every male even though there had been equal numbers of both sexes just twenty years earlier.

This sex imbalance may be at least partially explained by the fact that males are slightly smaller and have slightly shorter bills than females, so they're able to reach even fewer *Loripes* clams and are likely even less well-nourished and able to survive than females. Researchers are also finding fewer male than female chicks hatching in Siberia, though the reasons for this are not fully understood.

Because Red Knots are monogamous, mateless females are unable to breed. They will fly the dangerous course between the Arctic and Africa again and again, and produce no young. We are watching as the Afro-Siberian Red Knot, this branch of a species that has survived tens of thousands of years of climatic ups and downs, heads into extinction within our lifetimes.

# 5

# Fire

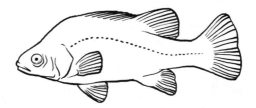

Mannus Creek is a small stream that rises in the rural highlands of southeast Australia near the insignificant village
of Rosewood, about halfway between the metropolises of Sydney
and Melbourne. From its source it runs for about twenty miles
through bland, mid-green cattle paddocks before flowing under
a fence line into the Bogandyera Nature Reserve, which protects
over twenty thousand acres of native eucalyptus forest, a remnant
of the diverse ecosystem that covered this region before ranchers
leveled millions of its trees and replaced them with a few species
of imported grasses for cows.

The stream follows a winding course through the gentle,
rounded hills of Bogandyera, wearing occasional deep pools into
its hard rock bed. About halfway across the reserve it flows steeply

down a sloping sheet of granite, Mannus Falls, and then, for six miles or so below the falls, becomes home to a small, valuable population of grumpy-looking fish—the last group of Macquarie Perch for hundreds of miles around.

Macquarie Perch are dark-colored predators that can grow almost twenty inches long, weigh over seven pounds, and can live for over twenty years. They have narrow, high-arched bodies generously rimmed with rounded fins and fronted with oddly small heads. They use their large, round bright-silver eyes to spot small fish and other aquatic animals that they suck into their low-slung, large-lipped mouths.

Adult Macquaries maintain small territories, often in larger pools that allow them to escape from high summer temperatures by swimming down to cooler, deeper water. Every southern spring, around early October, when the water temperature rises over sixteen degrees Celsius and river flow increases with rain, the fish will leave their territories. They'll swim up to several miles to congregate at traditional spawning riffles, stretches of river where the water speeds over shallow rocks. Here males will shed clouds of milt over the tiny eggs expelled by females, and their fertilized eggs will flow a short way downstream until the current slows and they drop down into narrow spaces between riverbed cobbles, where predators are less likely to find them. The fry will hatch in relative safety and begin to grow as the parents swim back to their habitual haunts.

Macquarie Perch used to be abundant in cool, fast-flowing streams in southeastern Australia. For thousands of years they were caught and eaten by Indigenous Australians, likely including the Bogandyera clan of the Wiradjuri nation, the people who lived in this area when European colonizers arrived. But from the mid-1800s white people began to change the fish's environment in radical ways. They introduced a distant relative from northern Europe called the Redfin Perch to the Macquarie's streams, and the Redfins spread a virus through many Macquarie populations and ate their way through vast numbers of Macquarie young. Colonizers built barriers such as weirs and dams across waterways, cutting off Macquaries' annual breeding treks. They deforested and plowed the adjacent land, turning clear rivers brown and smothering fish-spawning areas in layers of sediment.

In the early 1900s anglers could still pull tons of Macquarie Perch from well-known fishing spots inside a week, but by the 1970s the species was extirpated from much of its previous range and rare in the few rivers in which it still survived. Conservation efforts helped a handful of populations survive, but by 2010 Mannus Creek held the last population of Macquarie Perch in the New South Wales section of the giant Murray River basin. Fewer than a hundred adult Macquaries survived and bred in less than six miles of Mannus Creek downstream of Mannus Falls. Above the falls, a dam has made the water unsuitably cold, and introduced Redfin Perch prevailed. About six miles down from the falls, the creek is out of Bogandyera and flowing through cattle pasture again, its water quality is poor, and the Macquaries were outcompeted by non-native fish, including Asian carp.

The Macquaries held on in their Mannus Creek sanctuary even as southeast Australia entered a severe drought in 2017. Streamflow waned, and the Bogandyera reserve's forest slowly became drier. The drought continued through 2018 and 2019, but still the fish endured. As the Australian winter warmed into spring and summer, unusually large fires began to break out across the southeast of the continent.

In December 2019 extraordinary heatwaves persisted for days across this area, punctuated by occasional thunderstorms, and on December 29 a lightning strike sparked a blaze in the Woomargama National Park, about thirty miles away from the Macquarie Perch of Mannus Creek. At first it was a fairly modest fire, but the next day, December 30, it ran rampant. Extremely high air temperatures and rushing winds pushed it east into the Green Valley, toward the small agricultural town of Jingellic. Its flames rose hundreds of feet into the air, and its smoke went over forty thousand feet up—higher than jetliners' cruising altitudes. Its smoke plumes created so much friction that they generated their own lightning. At times giant, swirling vortices grew up out of the blaze, whipping up winds of over two hundred miles per hour. Just outside Jingellic they swept up and overturned a ten-ton firetruck, killing one of the fire-fighters within it. Within twenty-four hours the Green Valley–Talmalmo fire, as it was now named, burned across a quarter of a million acres of natural bushland and farmland.

The next day the air temperatures and windspeeds dropped, but the fire wasn't nearly done. It was so large that it effectively produced its own weather, lifting bits of burning material so high into the air that they came down and started spot fires over ten miles in advance of the main fire front. It consumed its way relent-lessly eastward, and on January 7, 2020, it entered Bogandyera.

The fire worked across the nature reserve diligently, combusting almost everything it touched. The flames went all the way to the edge of Mannus Creek, burning the riparian vegetation down to nothing, leaping the narrow waterway with ease. Inside two days almost the whole reserve had been thoroughly burned, the grasses and underbrush gone, the trees shrunken into skeletal black forms, only the trunks and largest branches left standing.

Huge numbers of animals had died, including deer and Eastern Gray Kangaroos. The animals fled into cool, moist gullies along-side the creek, shady sanctuaries filled with ferns, as the flames advanced. But this fire didn't skirt the gullies as other, regular fires might have; it worked its way around and inward, turning all the fine plants to hissing, pale smoke, leaving the desperate animals backed up together, grasping each other, dying in tangled heaps.

Bogandyera's land and most of its terrestrial creatures were burned, but the waters of Mannus Creek and aquatic life were relatively untouched. Yes, a few smaller animals like tadpoles and insect larvae in shallow water had been pushed above their temperature limits and killed as the fire's radiant energy quite briefly entered the water, and everything was now much brighter under the surface, with the riparian vegetation and its shade gone. But the stream flowed on in the days after the blaze, providing a relatively stable home for its inhabitants, as it always had.

———

What we call fire is a common outcome of combustion, a high-temperature chemical reaction that releases energy to the surroundings. Combustion requires two fundamental ingredients: a fuel and an oxidant. In nature the fuel is often plant material—wood, for example—and the oxidant is atmospheric oxygen. Combustion that is proceeding at such a rapid rate that it produces a visible flame is fire. (Wood can also combust slowly by rotting, which is decomposition by bacteria, fungi, and so on.) It releases a lot of thermal and radiant energy into its surroundings and generates a mixture of mostly gaseous products we call smoke.

About half of the content of wood is cellulose, a hydrocarbon molecule that can form long chains and is the primary structural component of plant cell walls. Plants make cellulose from the sugars produced by photosynthesis. Cellulose begins to burn when enough thermal energy from an ignition source is added to it to raise it to a temperature of 350 degrees Celsius (662°F)—if you hold a burning match to it, for example—which makes it begin to undergo thermolysis, chemical decomposition caused by thermal energy inflow (heat). The long, complex cellulose molecules vibrate more and more intensely and break down into various gases, aerosols, and solid char, and many of these then combine with oxygen in the air to form carbon dioxide and water vapor.

Thermolysis (the initial decomposition of cellulose into smaller, simpler molecules) requires the input of thermal energy from outside—it "uses" energy—but when these smaller molecules combine with oxygen from the air, a lot of thermal energy is released. This released thermal energy can cause more cellulose to decompose, releasing more of these smaller molecules, which oxidize and release more thermal energy—a chain reaction.

Fire is thus a chain reaction that only ends when the fuel (in this case, wood cellulose) or the oxidizing agent (atmospheric oxygen) runs out, or if not enough energy is released and reapplied to the fuel to sustain the chain reaction.**

As anyone who has tried to start a campfire in the rain knows, it takes a lot more energy to get wet wood to burst into flame than dry wood. And as we all know, dumping a bucket of water on a campfire is a good way to stop it burning. Liquid water suppresses fire because liquid water molecules can absorb a lot of thermal energy; if you hold a burning match to a wet piece of wood, much of the energy from the match will be absorbed by liquid water molecules and directed away from cellulose molecules in the wood, so they will take much longer to reach 350 degrees. For the same reason, adding water to a fire massively reduces the amount of energy available to maintain the chain reaction. Water can also form a physical barrier between fuel and oxygen, preventing the oxidation reaction.

This is why fires start more easily in drought-affected vegetation, with a lower-than-normal water content, than in well-hydrated

---

** Fire is, in a sense, rot sped up. Both processes break down cellulose and the other components of wood to produce carbon dioxide gas, water vapor, and other products; they both reverse the process of photosynthesis but have very different effects on the ecosystem.

vegetation, and why they burn hotter and spread faster during drought; water in both living and dead fuel is the most important constraint on fire in natural habitats.

———

Rain, blessed rain.

It's January 19, 2020. Eleven days ago the Green Valley–Talmalmo fire burned around Mannus Creek and across Bogandyera. There has been no rain since the fire, but now there's a radical change in the weather. Large thunderclouds have built up, and cascading sheets of drops are splashing on the soft ash that covers the earth, driving water down into the powdery remains of flowers, leaves, branches, tree trunks, insects, flesh, fur, and feathers.

Ash plus water is a promise of new life. For millions of years this mixture of vital nutrients and liquid has nourished plant regrowth after fire. But today is different. In many parts of Bogandyera, the fire burned so hot that it consumed the organic matter within the soil, leaving just sand and ash, which the wind has blown about, clearing it away from some places and piling it up into small, gray berms in others.

Across much of Australia, plain sand is hydrophobic; it resists water, casts it off like the feathers on a proverbial duck's back. So as the rain comes down it runs off the sand and soaks into the ash, turning the latter into softer and sloppier mudlike mush, which slides off the hillsides and into Mannus Creek. If it continues raining after the ash is swept away—as it does now—it becomes the turn of tons of sand and rock, with no vegetation to hold it, to wash into the stream.

As ash enters the creek, molecules within it oxidize, consuming dissolved oxygen in the water. The ash also fertilizes bacterial, algal, and fungal blooms, which use up even more oxygen. All along Mannus Creek, suffocating aquatic creatures are now trying to escape. Small crayfish swim to the surface and crawl out of the stream, tadpoles try the same, and even larval dragonflies break out of their skins early in a desperate attempt to become adults and fly away; their wings are tiny and misshapen, and they can't lift off.

Fish, including Macquarie Perch, open their mouths wider and pulse their gill covers faster and harder to move more water across their gills to absorb what little oxygen remains. Ash and fine sand rush in, clogging up and abrading the delicate gill tissues, which begin to bleed. The fish float to the surface, lips gasping at the air, and slowly die. Even air-breathing Duck-billed Platypuses are forced out of their normal daytime refuges (they're usually nocturnal, and hide during the day). They swim about on the surface, clearly distressed, moving up and down their darkening, thickening pools.

Within hours the entire creek is transformed into a stream of blackish sludge with an oxygen content of close to zero. Deep pools fill to the brim with sediment and clumps of dead and dying aquatic animals are pushed downstream.

---

Terrestrial green plants can trace their origins to marine green algae species that evolved to live in fresh water and then emerged onto land as tiny, crude spore-bearing organisms about 470 million years ago. Not until about 430 million years ago did they begin evolving shoots, leaves, and the sort of internal vascular structure that we see in contemporary land plants, a process that allowed an explosion of many new forms and species that spread across the planet.

Lightning, which can ignite fire, has flashed inside the atmosphere for billions of years, but it wasn't until land-plant photosynthesis raised atmospheric oxygen to very high levels that fire became common. Although the earliest known fossils of charred plants have been dated to about 420 million years ago, it's likely that large fires that burned across extensive areas for extended periods of time became common only about 60 million years later, in the early Carboniferous period. By then large, woody trees had evolved and formed forests across many parts of the globe, abundant fuel for blazes.

But even given oxygen-rich air, fire couldn't—and still can't—burn through all types of vegetation. Some vegetation types grow in areas that are simply too wet to sustain fire. Plant species that evolved in wet habitats have never had to develop the ability to survive it. If they're exposed to flames, they usually die. But many plant species that evolved in drier areas have developed various means to survive fire.

Fire, therefore, can be a strong determinant of which species can live where, sorting species into separate ecosystems according to their fire sensitivity. In Australia this process has divided the continent into a large "dry world," the fire-prone and fire-tolerant ecosystems that occupy the vast majority of the continent, and a far smaller rainforest "wet world" that exists in a narrow north-south band just inland from the east coast.

Australia's dry and wet worlds have very different evolutionary histories. The large dry world is dominated by relatively newly evolved species, but the wet world contains many notably ancient lineages, some of which evolved on a very different planet than today's and have remained almost unchanged for tens of millions of years.

The story of Australia's wet world is a remarkable one of escape and survival over enormous distances and deep time, and we can begin it about fifty-six million years ago, at the beginning of the geological Eocene epoch.

The Eocene dawned when the average global temperature rose by about six degrees Celsius in just twenty thousand years, probably triggered by large volcanic eruptions releasing enormous volumes of carbon gases. Significant extinctions resulted: fossil remains show that between 30 and 50 percent of marine benthic foraminifera (seabed-dwelling single-celled organisms) went extinct.

Continental landmasses were arranged very differently in the early Eocene than today, shaping quite different patterns of ocean circulation: the Isthmus of Panama had not yet linked North and South America, so water flowed freely between those continents. The southern tip of South America was still joined to Antarctica, and Australia had only just begun to break away from the southernmost continent and move northward. (These three continents had previously been joined together with Africa and India as the supercontinent of Gondwanaland.)

The southernmost landmasses were warm and humid, a result of extremely high atmospheric carbon dioxide levels. The poles had been partly or completely ice-free for tens of millions of years, giving plenty of time for forest to colonize them and evolve to the conditions, which included over two months of permanent daylight in the summer and a similar period of darkness in winter. Plants photosynthesized continuously during the summer, building up carbohydrate stores to consume during the winter, when they could not photosynthesize. (An early dinosaur that lived around the South Pole about one hundred million years ago, *Leaellynasaura*, which was about the size of a small kangaroo, had enormous eyes, presumably so it could see in the dark months.)

As the Eocene unfolded, South America broke away from Antarctica, forming the Drake Passage. Water could now flow continuously around Antarctica, which gave rise to the Antarctic Circumpolar Current, which still today runs clockwise around the southernmost continent. Global average temperatures gradually dropped; about forty-five million years ago, ice began to persist year-round in parts of Antarctica.

Since cold air holds less water vapor than warm air, the Eocene temperature decline led to lower and lower global average rainfall. Australia continued moving north, toward the equator, exposing it to relatively warmer temperatures. It began to aridify, its vast interior—farthest from moisture-producing oceans—drying out first. It began losing its Gondwanan rainforest plants from the center outward; they were replaced with a growing and rapidly evolving drought- and fire-tolerant flora.

Australia's dry world was coming into being, radiating from the center outward, and the wet world was losing ground, not only on this continent but on Antarctica, too. By the end of the Eocene, about thirty-four million years ago, Antarctica was largely covered in ice, which killed off most of the Gondwanan plants. Although South America had already separated, it was still near Antarctica; its southern reaches froze and it lost almost all of its Gondwanan forests too.

One area was kind to the ancient plants and some of the animal lineages that had evolved with them around the South Pole: a massive complex of mountain ranges, peaks, and uplands running just inland, almost the entire length of the eastern side of Australia, which today is called the Great Dividing Range. These high-altitude areas created consistent year-round rainfall by forcing the humid air that pushes westward off the Pacific up to higher altitudes, where the water vapor within it condensed into clouds, creating regular rain in the mountains. This resulted in a series of refuges with climates similar to that of the ancient Gondwanan landmass of tens of millions of years previously, sanctuaries that today stretch from the island of Tasmania all the way to Cape York, the northeastern tip of Australia.

Today less than 1 percent of Australia is covered with rainforest, but this tiny area is home to about half of the continent's wild species, including about 60 percent of its plants. Walking into the Gondwanan forests of the eastern Australian highlands is like a trip back in time; here you can see groves of fat, gnarled, moss-covered *Nothofagus* trees rising from a forest floor covered with ferns very similar to those that the dinosaurs ate. *Nothofagus* is called Antarctic Beech by contemporary Australians, but it's not related to the beech trees of the Northern Hemisphere; its leaves, bark, and internal wood structure are virtually identical to sixty-million-year-old fossils found in Antarctica. (Australian tree common names are often inspired by a species' timber. If the wood color and structure is similar to a well-known species in the Northern Hemisphere, it often gets named similarly.)

*Nothofagu* is not the only ancient tree species on the wet strip of eastern Australia; in 1994 botanists discovered a small grove of fewer than a hundred massive, old, and very strange pinelike trees in a wet gorge in the southern Great Dividing Range. At first they didn't know what they were, but careful examination of their pollen showed them to be an ancient lineage that first showed up in ninety-million-year-old Gondwanan fossil beds, a primitive species distantly related to the also-primitive *Araucaria*, or Monkey-Puzzle trees, that also still grow here. They named the brand-new ancient species *Wollemia nobilis*, the Wollemi Pine. Even more recently, in 2000, a forest ecologist recognized that a group of about a hundred strange trees in a temperate rain-forest patch in northern New South Wales belonged to another ancient lineage, a primitive form of the proteas found today in Australia and South Africa, surviving by the tip of a proverbial fingernail after an achingly slow journey of thousands of miles over tens of millions of years. Most contemporary proteas are small shrubs, but this is a towering tree. He named it *Eidothea hardeniana*, the Nightcap Oak—even though it isn't at all related to the familiar oaks of the Northern Hemisphere.

Even the soundscape of these forests contains ancient elements; the songs of some of the most primitive surviving songbirds and amphibians on Earth. The Superb Lyrebird struts around the forest floor like a lithe, long-tailed pheasant, turning over leaf litter in search of grubs and perfectly imitating a huge variety of other sounds in the forest—including, in some areas, cell phone ringtones and chain saws. Despite its appearance, it's more closely related to sparrows than pheasants; it's one of the world's largest and most ancestral songbirds. The extraordinary, artistic Satin Bowerbird is another ancient species. Male Satin Bowerbirds collect small blue items to carefully arrange around a bower—an open arc of interwoven twigs—on the forest floor, in which they dance to attract mates. Some of the world's most primitive frogs are found here, too.

Australia's old wet-world lineages didn't have to change that much to survive; they only had to be lucky enough to find themselves in areas where a climate similar to that of ancient Gondwanaland persisted through time, or be able to move to areas with such climates. Their survival has been dependent on *where* they are less than *what* they are.

Because the dry world was new at the dawn of the Eocene and has been growing since, evolution proceeded relatively rapidly here. As Australia dried, some groups of plants evolved ever-better reproductive strategies and physical forms to thrive in water-restricted ecosystems and survive the fires that were increasingly able to spread in their habitats.

Plants adapt to survive drought in many ways. Some radically slow their metabolism during dry periods to avoid having to use, and thus lose, water. Dropping all leaves in the driest parts of the year is one way of doing this. Other plants go to extremes to avoid running out of water during low-precipitation periods, by growing extremely deep roots that can tap into groundwater, for example. And yet others have developed complex physiological mechanisms that allow their cells to survive even when their internal water levels are extremely low; they "dry out" and go into states of suspended animation, coming back to life when water becomes available again. Many plants have physical characteristics that lower their rates of water loss, like the so-called sclerophylls ("hard leaf," in Greek), with their hard- or thick-skinned leaves and small stomata.

Dry vegetation invites fire, so—except in hyper-arid desert areas, where there is too little vegetation to carry fires, and/or where many of the plants are water-dense succulents, which don't burn—plants have also evolved to resist it. As the dry world expanded across the continent, many Australian plants evolved not just to survive fire but to benefit from it. They became pyrophilic.

The eucalypts, or gum trees, are Australia's most iconic dry-world plants. The Blue Gum, *Eucalyptus globulus*, which is grown in plantations and in parklands around the world, is typical of the larger eucalypts. A few early eucalypt fossils have been found in old Gondwanan deposits in Patagonia—they appear to have been present in small numbers in drier parts of the humid Gondwanan world—but these plants truly grew into themselves as Australia dried out, diversifying into the over 750 species found there today. Some are among the world's tallest trees. Others are small shrubs. Yet others, known as mallees, form shrubby thickets with multiple stems sprouting from a woody, rootlike structure called a lignotuber.

Many eucalypts are drought resistant, with sclerophyllous leaves and deep taproots, but they also have adaptations that allow them to manipulate fire for their own benefit.

Eucalypts have oil-rich leaves, and many shed their bark as they grow. As the bark ages, it falls off in large, curled, dry strips that, being highly acidic, don't rot easily and make excellent fire fuel that accumulates over years around the base of the tree. These characteristics may seem suicidal in fire-prone areas, but there's method to this apparent evolutionary madness: the same trees that carry ultraflammable leaves and spread enormous piles of kindling at their feet also have trunks with very thick, insulating outer layers of living tissue. Deeply embedded in these layers are dormant epicormic buds—embryonic shoots. The trees' fruits are also small and hard, saturated with sticky, waterproof resin, and can remain on the trees for years, thus keeping them safe from seedeaters like birds and insects that might consume them on the ground.

When a fire moves through a eucalyptus woodland, the piles of dry bark on the forest floor combust hot and fast, burning away the forest floor vegetation and sending flames high into the oily leaves, which burn up even faster. The resulting pulse of released energy does not endure long enough to set the trees' trunks on fire, but it does melt the resin holding the seeds in the fruits. Shortly after the fire has peaked, the seeds rain down on a forest floor now cleared of competing plants and fertilized with ash. The loss of living leaves at the crown sends hormonal signals down the trees, which awaken the epicormic buds. The trees send nutrients from their storage roots up the trunks, feeding the rapid growth of new shoots. Within weeks, fast-growing sprays of bright-green leaves are popping out all over tree trunks, many of which will become new branches to replace those lost to the fire. Thus eucalypts promote the type of fast-moving fire that can prompt their own resprouting and eliminate competition for their seedlings.

Other Australian dry-world plants have different survival methods. Plants in the highly diverse *Acacia* genus—which consists mostly of shrubs and small trees—produce huge volumes of hard, water- and rot-proof seeds, starting at a young age (year-old seedlings can flower and make seeds). They drop these onto the soil around them, building deep seed banks in which seeds can remain dormant and viable for well over a century. Adult acacias of most species don't live long—twenty years is a lot for most species—and are typically killed outright by fire. But a pulse of heat and exposure to certain chemicals in smoke awakens their seeds and triggers them to germinate. Add a bit of rain, and dense monocultural stands of young acacias grow up, shading out competing species.

Grasses (Australia has many types!) have other techniques. Many perennial species build up nutrient stores in their roots and embed their growing shoots as deep within their bases as possible. They sacrifice their aboveground parts to the flames, their leaves becoming fertilizing ash, and regrow from their protected shoots. Some individual grass plants can live for over five hundred years, enduring hundreds of fires. Many annual grass species make seeds that can be widely dispersed by birds or the wind, so there are always some members of their species germinating and growing in a place that doesn't burn that year.

As there is no single way of surviving fire, there is also no single type of fire. Fires come at different intervals. They burn at different temperatures, move through ecosystems at different speeds, burn along the ground or high up, through the crowns of trees. Different ecosystems tend to encourage different types of fires—which in turn encourage different types of ecosystems. Definable fire regimes emerge from this interplay; as places have climates, fire-prone ecosystems have fire regimes.

Since water is such a strong controlling influence on fire, most fire regimes are structured around rainfall. Fire generally occurs in the dry season, which can be winter or summer, depending on the region. But fuel availability and the dominant type of ignition source also play large roles. Before humans started using fire on Earth, lightning was the most important fire starter, and regions that experienced numerous thunderstorms tended to have more frequent fire. Other natural sources of fire ignition include rock-falls (which generate sparks), volcanic activity, and rotting vegetation (which can generate remarkably large amounts of thermal energy).

Regular lightning can increase the frequency of fires, creating a short-turnover fire regime, but only if there is enough suitable fuel to carry a fire after ignition. It can take years after a fire in a eucalyptus-dominated woodland for sufficient fuel to build up so as to allow a fire to burn through it again; bark must grow, fall off the trees, and form a thick enough layer on the ground. Eucalypts also shade the ground, which excludes fast-growing grass species, further increasing the time between fires. Grasses can regrow rapidly after fire and create sufficient fuel to burn again in less than a year, so grasslands commonly burn more frequently than other vegetation types.

Vegetation can exert strong influence on fire frequency and vice versa. Particular vegetation communities and fire regimes form in response to each other and reinforce each other, creating predictable, repeating patterns in space and time—until something larger, like climate breakdown, throws one or both elements of the system off track.

---

For hundreds of years fishermen and sailors in the Pacific Ocean off the coasts of Ecuador and Peru have known about a warm current that sometimes flows south along the coast, usually in December. They call it El Niño, "the Boy," in reference to the Christ child, whose birth they celebrated around the same time. In some years the temperature of El Niño is notably higher than normal, and then the weather changes inland; massive storms form over the ocean and travel inland, dumping colossal volumes of rain and causing damaging floods.

Since El Niño was first named, scientists have learned that its influence extends far beyond part of South America, and that it's just one aspect of a far larger system. They've identified a large, irregular periodic shift in the sea surface temperatures and air pressures above the eastern tropical Pacific Ocean, nowadays called the El Niño–Southern Oscillation (ENSO). Careful, long-term observations have found that every two to seven years, the surface waters in this area become significantly warmer than usual, changing major weather patterns in the tropics and subtropics around the world. In these warm El Niño years, rainfall increases in western North and South America but decreases in China and eastern Australia. Northeast African rains increase, while southern Africa goes into drought. When the surface water in the tropical eastern Pacific drops lower than normal, these patterns usually reverse.

But the ENSO isn't the only massive climatic pattern determined by surface water temperature in patches of the ocean. In recent years meteorologists have defined a phenomenon called the Indian Ocean Dipole (IOD), which is defined by the difference in sea surface temperatures between two areas ("poles"), the first located in the Arabian Sea in the northwestern Indian Ocean, and the second being south of Indonesia in the Indian Ocean. A "positive event" in the IOD occurs when the water temperature south of Indonesia is higher than that in the Arabian Sea. This causes changes in bulk air circulation over Australia, reducing rainfall in the south and east. The greater the temperature difference between the poles, the stronger the effect. As for El Niño, a positive IOD event usually occurs in a single year, flipping back to negative afterward.

In 2018 an El Niño and positive IOD event occurred simultaneously and persisted for months. The rains failed over eastern and southern Australia, as would be expected from past observations.

And then in mid-2019 this doubling up happened again, even more strongly than the previous year. Scientists saw the signature of climate change in these unprecedented events.

By late August—the end of the usually wet Australian winter—large fires had broken out across the south and east, and they kept starting and burning all through the spring, torching hundreds of thousands of acres, an unprecedented situation for that time of year.

Summer brought no respite. A massive high pressure dome hung over the southeastern part of the continent, keeping moist ocean air away. December saw hundreds of local temperature records shattered, some places in the southeast—which is not a desert—exceeding 43 degrees Celsius (110°F) for two weeks. Tens of millions of acres of soil dried out, followed soon after by the plants growing in that soil.

With the vegetation so dry, fires burned hotter and faster than in all of Australia's recorded history, intense enough in many places to overcome the deeply evolved defenses of eucalypts, which were no longer holding enough water in their trunks to confine the flames to their leaves. Masses of thermal energy penetrated and split their living tissues, setting heartwood aflame. This rich, normally-off-limits fuel burned hotter and longer, setting more trunks alight, liberating so much energy that—as in the Green Valley–Talmalmo fire that began this chapter—columns of smoke regularly pushed over twenty thousand feet into the air, generating pyrogenic dry thunderstorms that threw down lightning bolts miles in advance of the fire front, casting flames before flames, fires amplifying into megafires—unimaginable, but real.

These fires burned thoroughly, leaving almost no patches unconsumed, none of the refuges where small animals like mice, lizards, or butterflies find sanctuary in regular fires. Even flying birds burned; at various places on the southeastern coast people saw flocks of parrots screaming away from the smoke, their wings and tails aflame, heading out over the ocean. With nowhere to land, they drowned, and their charred corpses washed ashore in the days after. Many fires burned so fast and from so many sides that even large kangaroos couldn't outrun them. Overcome by smoke, these strange, graceful animals collapsed and boiled inside their own skins. Tens of thousands of slow-moving Koalas died in agony in just a few weeks. The skies turned a constant pale gray for hundreds of miles around, with deathly deep-red sunsets day after day after day.

And the wet world, the ancient forests of the Great Dividing Range? Many were no longer wet. For the first time in thousands of years, in some places maybe for the first time ever, fires came in around their edges and worked their way in, often thousands of acres in. The flames cleared out the understory, the delicate ferns, the soft, green-leaved shrubs, and then attacked the trees, big and small. Many centuries-old giants fell as they burned, their trunks turning to giant smoking logs eating into the soil, their hearts glowing red for weeks before expiring as long lines of charcoal, fallen shadows on the forest floor.

By January 2020, over sixty million acres of Australia had burned, primarily in the southeast. The fires almost doubled the country's average carbon gas emissions, and their smoke traveled more than seven thousand miles across the Pacific, dulling skies in Chile and Argentina. About four billion animals and birds probably died, too, including a third of the entire population of Koalas.

---

Some people say that if you push your nose into one of the deep, dark cracks in the reddish-yellow bark of a big old Ponderosa Pine and take a deep breath, you'll smell vanilla. Others say you'll smell something like butterscotch, chocolate, pineapple, or turpentine— or perhaps nothing at all. But before you nuzzle your snout into the tree, look carefully in the crack; it might be the home of a venomous spider or a biting insect of some kind.

Ponderosas grow across millions of acres of the American West, favoring mid-altitude mountains and plateaus from northern Mexico all the way to southern British Columbia. They're the largest organisms in many areas of their range, growing up to nearly 270 feet tall and over 20 feet in girth, and living up to a thousand years. They have a single, straight trunk from which horizontal branches grow, covered in bunches of thin, shiny green needles up to seven inches long, and large prickly seed cones up to five inches long. Ponderosa seeds are small and light, with a small wing-like projection. When cones mature, they release seeds that can be carried some distance on the wind, to germinate away from the crippling shade of their parent trees. Young Ponderosas have blackish bark that turns yellowish-red at about a hundred years of age. Their wood is heavy—ponderous—and yellow. Before the late 1800s, when masses of European people began moving into and fundamentally changing the American West, Ponderosas formed vast, grand, parklike woodlands. They usually grew far apart, giving meadow grasses and wildflowers the light and space to abundantly carpet the earth around them.

In many places where Ponderosas grow, they're a keystone species, strongly defining and shaping the appearance, structure, and functioning of their ecosystems. They provide food and habitat for numerous mammals, birds, reptiles, arthropods, fungi, and other plants. Tassel-eared Squirrels, found in the southern part of the Ponderosa's range, are dependent on the tree and occur only with it. Tassel-ears are midsize squirrels, gray with a red-brown blaze down the middle of their backs and inch-long fur tufts on the ends of their ears. They not only live in Ponderosas and eat their seeds but also form an important ecological triad with the tree and a fungus called the False Truffle.

Most Ponderosas grow in low-nutrient soils, and like many other types of pine trees they've evolved a mycorrhizal association between their roots and an underground fungus. The tree supplies the fungus with sugars and carbohydrates produced by photosynthesis, and the fungus efficiently draws out nutrients—like phosphorus—from the soil, and passes them into the tree. The False Truffle grows whitish, roughly spherical fruiting bodies underground—about an inch in diameter—that generate its powdery, gold-hued spores. Tassel-eared Squirrels find False Truffle fruiting bodies tasty and dig them up, scattering the spores on the soil surface, where they can blow away, dispersing neat, spore-filled golden droppings that further spread False Truffles through the ecosystem.

Lightning is common all across Ponderosa territory, and because of their height, the trees often act as lightning conductors. Walk along a ridge in a Ponderosa grove, and you'll often find trees that have been struck but are still alive and growing, with a line of split-open bark running from crown to roots or cavernous, blackened holes burned into the heartwood. Before European settlers came, lightning-ignited fires often burned through Ponderosa groves, but they killed few mature trees: the Ponderosas evolved to benefit from the fire regimes common across the American West before large numbers of white settlers moved there; that is, frequent, low-intensity fires that traveled along the ground.

Mature Ponderosas have thick, insulating bark, and as they grow, their lower branches die and drop off. If lightning struck a Ponderosa in a natural grove and set it alight, the flames would likely not spread to other trees, because they grow so far apart; flames would move through the grass layer along the ground. Because grass burns quickly, these fires would move past trees rapidly, without releasing thermal and radiant energy for long enough to set their trunks alight. As Ponderosas mature, they drop their lower branches, removing woody "ladders" that can carry flames from the ground to a tree's living crown.

In the past, most fires in Ponderosa country would stay small and on the ground, killing very young trees (Ponderosas and other species) that compete with older trees, and spreading nutrients in the form of ash on the soil surface. Since Ponderosas are so long-lived, very, very few of their young need to reach maturity to sustain their numbers, so the loss of most small trees to fire doesn't affect natural Ponderosa populations.

But this well-functioning relationship with fire was upset when Europeans arrived in the American West. Millions of large old Ponderosas were cut down for their high-quality timber, and vast acreages of forestland came under the management of foresters with European training, to whom fire was anathema. (The few white foresters who knew of Native American burning practices considered them primitive and harmful.) The newcomers thought that the best way to maintain an abundant supply of wood was to suppress fire and allow timber trees—including Ponderosas—to grow closer together. What use were grasses and wildflowers to the burgeoning capitalist economies rising up on the West Coast?

Ponderosa saplings began to grow up in tightly bunched, uniform stands after loggers had passed. Fires were enthusiastically suppressed—whenever flames broke out, government firefighters were soon on the scene. Smokey the Bear became the icon of forest conservation. Over decades, western forests became denser and darker. Then, in the 2010s, came drought. The rains failed, year after year. The forests became tinderboxes, and when fires came, they had so much fuel that they became uncontrollable crown fires, tearing up the entire height of the trees, killing them.

Some Ponderosa seeds can survive intense blazes. After fires have passed, they germinate and begin to grow in ground that in many areas hasn't experienced direct sun in over a century. It might seem like a chance for the Ponderosa ecosystem to reset itself and reform, but it's not. As with most places around the globe in this age of climate breakdown, maximum air temperatures across the American West have risen just enough to drive a critical amount of moisture out of the drought-parched soil across much of the Ponderosa's range. Many young Ponderosas die from drought stress. In other places, direct sun raises the soil surface beyond boiling temperatures, killing the living tissues of saplings at the point where they exit the earth, girdling the young plants.

The Ponderosa Pine is not the only species of large conifer that forms extensive forests in the American West. Many others do, too, including Jeffrey, Lodgepole, Piñon, and Whitebark Pines. These often react similarly to fire, which means that trees are not coming back across millions of acres of the American West that have burned in recent years. Superhot megafires have killed nearly all the mature trees they've raced through, and tree seedlings cannot survive in the warmer, very much more extreme world that's emerging. Acre by burned acre, these ecosystems are flipping into low scrubland, dominated by fast-growing shrubs and grasses. And jays, woodpeckers, chipmunks, squirrels, hawks, owls, bats—every species that needs Ponderosas and other large trees to survive—are disappearing too.

---

Australia's Black Summer bushfires ended when the high-pressure system that had hovered over the continent retreated, allowing a series of large rain-bearing fronts to move inland and douse the flames. Shortly afterward the Indian Ocean Dipole reversed polarity, flipping southeast Australia into a wet annual cycle.

This would normally bring relief to burned ecosystems and allow them to regenerate. But this has not been the case for many burned regions. Thanks to unprecedented increases in ocean temperatures, the Indian Ocean Dipole's reversal has been longer-lasting and stronger than any on record, and rainfall records have been broken across southeast Australia in each of the three years since the fires died down. In 2022, some regions received more than five times their average annual rainfall in just a few months. The rainy season has become an extended flood season, ripping millions of tons of soil and dirt from the floors of burned forests and dumping them into rivers, streams, estuaries, and the ocean. Pools have filled up, riverine vegetation has been torn away, and hundreds of miles of river have turned from clear to permanently muddy. The famed Duck-billed Platypus is vanishing from about half the rivers in its range, starving because it can't feed on sediment-laden river bottoms. Massive algal blooms, fertilized by huge new nutrient inputs, are turning estuaries anoxic and green. Flood-related fish kills are now common.

Vast areas of Alpine Ash, a type of eucalypt that forms extensive, dense forests in the Great Snowy Mountains, are not regenerating. The species cannot resprout after fire; it reproduces from seed. Alpine Ash trees only reach reproductive maturity and start making seed at about twenty years old, and their seeds remain viable in the soil for less than a year. If a large fire passes through a stand of Alpine Ash and kills all the trees and the seeds in the soil beneath them, the stand will not regenerate. Even if a few seeds survive and Alpine Ash saplings begin to grow, it will be at least twenty years before they can reproduce. With large fires coming more frequently, Alpine Ash is being eradicated from areas where it was dominant for millennia. It's being replaced by dense thickets of Acacia shrubs sprouting up from seed banks deep in the subsoil, many laid down over a century ago. Few birds sing, and fewer mammals and reptiles can be found in the new brush; the species evolved to live in Alpine

Ash forest can't live in this habitat with its completely different structure and species composition.

In the wet world, things are confusing. Some wet forests are going the way of the Alpine Mountain Ash, overtaken by sheets of *Acacia*. Other, more lightly burned groves are regrowing with wet-world species, but extremely slowly. In most fire-affected places nearly all the grand old trees were killed. The forest canopy disappeared with their demise, and now the forest is half its previous height—a dense, low thicket of regrowing shrubs and trees lorded over by the pale gray skeletons of the ancients, huge dead branches sticking up into the open air. Many mammals, like Koalas and Yellow-bellied Gliders (which are like large, marsupial flying squirrels), are gone. They burned in the flames, and there is no habitat left for the survivors. Some of the wet-world forests may recover their former form, but it'll take centuries.

These ecosystems have undergone a state shift; the disruption of the old fire regime by superhot megafire was so powerful that their traditional cycles have been disrupted. Rising temperatures and changes to age-old weather patterns have played havoc with these unique environments. They cannot bounce back from disturbance in the way that they evolved to. They're becoming something new.

Small amphibians like the tiny Pouched Frog, less than one inch long, have been hit hard. Pouched Frogs come from an ancient lineage; fossils of their immediate ancestors show up in thirty-million-year-old Antarctic rocks. They've lived on the constantly wet floors of Gondwanan forests for so long that they've evolved a weird breeding system. Their tadpoles don't live in water; adult males carry them around in small pouches in their skin, and the wet environment keeps the tadpoles constantly moist. With the forest canopy gone in the southernmost parts of their range, Pouched Frog habitats are drying out rapidly and the frogs are unable to breed successfully. Late 2022 surveys in areas that each held hundreds of Pouched Frogs previously found between zero and ten individuals today.

Some birds, too, are almost gone from areas where they thrived before Black Summer. Up in the wet temperate forests of the Nymboida River catchment of northern New South Wales are a few long river valleys lined with dense forests of she-oak trees, the tiny, seedlike fuits of which are the main sustenance of Glossy Black Cockatoos. These large parrots, black all over but for bright scarlet flares in their tails, can live for seventy-five years and mate for life, each pair producing only one youngster every few years. A recent survey found fewer than ten individual stragglers in one valley, the remains of a population of hundreds of birds. After the fires killed the she-oaks, the birds had starved. (She-oaks are regenerating here, but it'll take at least fifteen years for the trees to start producing seed.)

And Mannus Creek? It may take decades to become top-quality habitat for Macquarie Perch again, if it ever does. In early 2021 conservationists released hundreds of young, captive-bred Perch into the stream, heartened by the good postfire rainfall. They assumed that, as with fires in that region and habitat in the past, rain would gradually clear the sediment from the creek, hollowing out refuge ponds and clearing sediment from the fish's breeding riffles. But the extraordinarily high intensity and high volume of the rain in that year and 2022 has had exactly the opposite effect: it's washed ever-greater volumes of soil, sand, and rubble into the stream, replacing the black ash with red-brown dirt. There are few deep pools left in Mannus Creek, and the riffles remain largely clogged. Some Macquaries survive here, but it will likely take many years of regular, moderate rainfall to reform stream habitats as they were before the Black Summer fires. Regular, moderate rainfall is not on the horizon, though. Warming is likely to drive the Indian

Ocean Dipole between ever-stronger extremes and increase El Niño intensity, bringing drier droughts and stormier rainy seasons to southeast Australia.

Research shows that the global area of natural vegetation burned by fires is not increasing. Humans have fragmented natural landscapes with roads and farms, which make effective firebreaks that constrain the spread of blazes. But data show clearly that fires are becoming more intense almost everywhere, releasing more energy and burning up more of the systems they move through, extirpating species and changing ecosystem structure. As with hurricanes, intensity matters; the new megafires have completely different effects than the slower-burning, cooler fires of just a few years ago.

# Fertile Air

A Cheetah lies still and low, eyes forward, in a mass of small, scrubby thorn trees, mostly Sicklebush and Blackthorn, in the Otjozondjupa region of northern Namibia. She's facing into a lightly-grassed clearing criss-crossed by the habit-worn trails of Gray Duiker and Steenbok, small antelope she's become accustomed to killing.

Only one cub from her most recent litter survived to adulthood, and he left some months back to find a territory of his own, so she woke alone an hour ago. She indulged in a long, tongue-flexing, canine-baring yawn as the first pulse of real blue began to wash the stars from the eastern sky, and then picked a careful way through the dense, prickly bush to this small arena, two or three open acres of scrappy Wool Grass and Dubi Grass on reddish, sandy soil. A few sniffs and glances at fresh hoofprints told her that it was well enough used, so she settled down in ambush as the day's hard yellow light hit the branches above her hiding place and the birds rose around her into abundant, rippling morning song.

Twenty yards away, to the Cheetah's left, a male Gray Duiker stops at the edge of the clearing. With a foreleg frozen in midair, he silently jerks his dark, wet eyes and large ears toward an almost imperceptible rustle—toward her. Flicks and shivers pass along his skin, sending up a couple of early flies.

The Cheetah tenses up, her belly fur tickling across the loose sand, and then explodes forward. With no time or space to turn, the Duiker shoots into the clearing, the Cheetah turning hard right to intercept his headlong sprint across the space. She swipes at his back legs with an outstretched paw to trip him, but misses, then tucks in behind him for another try, her nose just a yard from his rear.

It takes just seconds for them to reach the opposite edge of the clearing, where the Cheetah suddenly breaks off her charge and stumbles to a halt, shaking her head in a thin cloud of dull, slowly rising dust. The Duiker disappears in a waning cascade of hoof thumps, safe again in the thicket. The birds have quit singing. An inch-long Sicklebush thorn, one of many on a long, thin, low branch, has ripped into the Cheetah's right eye.

In the coming days her eye will swell and its cornea will turn a pale powder blue. She will begin to starve, a loose-skinned Cyclops loping through the bush in search of easy meat—a calf on a nearby cattle ranch, perhaps, or the remains of another predator's kill—but she won't find it.

One evening, as she looks for shelter, she'll be surprised by another spotted cat: a glossy male Leopard with a dark mouth and bright teeth, the new lord of the thicket. There will be snarling and he'll attack, grasp her throat in his jaws, and wrench her life away. At half his weight and much slighter, she wouldn't have had a chance against him even if her eye was still good.

Her fate is no longer unusual. Although the Cheetah has thrived across vast swaths of northern Namibia and the surrounding regions for millennia, in recent years it's begun losing ground to the bulkier, slower Leopard, a species that used to be very rare in Cheetah territory. This dangerous newcomer is arriving with an extraordinary rush of woody plants—shrubs and trees—that is rapidly and radically remaking landscapes across southern Africa.

The Cheetah evolved from the same ancestral lineage as the Puma and Jaguarundi of the Americas, which are its nearest living relatives. But unlike those cats, which are highly adaptable generalist predators that live in a wide variety of habitats, the Cheetah lineage has consistently, over the past two and a half million years, evolved for a life centered on speed.

The modern Cheetah chases its prey farther and faster than any other cat species, and is in fact probably the fastest land animal that has ever existed; no known fossil species could plausibly have run at its pace.

Its adaptations for speed make it arguably the strangest cat of all. It has an oddly small, short head, low-slung in front of a slender body; an aerodynamically efficient package. Its heart and lungs are relatively larger than any other cat's, to move oxygen to its muscles more rapidly, and its vaulted braincase houses enlarged nasal passages to move huge volumes of air into the lungs and move body heat outward, fast. Its relatively small jaw and teeth save weight.

Unlike those of other cat species, the Cheetah's claws are not fully retractable, giving it better traction when running. Its footpads are hard and sharp-edged, providing better grip at speed. Its leg bones are long and thin, its spine extra-flexible, and its muscle structure optimized for running. Its long, heavy tail provides a steadying counterweight to its body, which is useful while sprinting and in tight turns. Any body mass that could slow it down has been evolved away.

The Cheetah's eyes are wide-set above a set of black "tear marks," to reduce glare in open sun, and it has a far greater proportion of short-wavelength light receptor cells in its retinas than other cats. This gives it poorer vision in shade and darkness, but an extraordinary ability to track fast-moving objects in bright conditions. Its ear vestibules are enormous, holding highly sensitive balance and motion sensors to keep its head unnervingly steady and locked on target during a high-speed chase.

While other cats have strong forelimbs, which they use to tackle and grasp their prey, the Cheetah's forelimbs are slender and comparatively weak, with light bones and refined muscles dedicated to forward motion, not grabbing. It uses an elongated dew claw, the inner "thumb" claw, which is set up off the ground, to hook and trip its prey. At the speed that it engages, it doesn't take much power to throw a prey animal off balance and send it tumbling into the dirt. The Cheetah quickly bites its throat and suffocates it.

These adaptations make the Cheetah a highly efficient predator. It notches up one kill for every two attempts, a higher success rate than other big cats. But it can't hunt just anywhere: it needs unobstructed terrain to sprint across. It needs good lines of sight, and good light, to home in on its fast-moving prey. It needs a safe place, where it can see approaching animals, to eat its kill; being weak, it can easily lose its meal or its life to other predators. The Cheetah's only real defense against Lion, Leopard, and Spotted Hyena is to run away.

The Cheetah has been able to develop its extreme adaptations to speed and become an extraordinary creature of space and light only because it, and its forebears, had consistent access to large open habitats as it evolved. This strange animal did not arise in a dense, dark forest.

---

In the early 1800s scientists of the type that we now call ecologists began trying to explain why particular forms of vegetation were found in particular places. Why do forests grow where they grow? Why grasslands? Why tundra?

The German polymath explorer Alexander von Humboldt, who was fascinated by the relationships between organisms and their environments, was an early pioneer in this realm of inquiry. On his trips to South and North America between 1799 and 1804 he developed the idea that vegetation types were related to local climatic conditions, and described different vegetation zones in relation to the latitude and altitude they were found at and the local temperature and humidity, a conceptual framework that eventually grew into the scientific field of biogeography. (He also attributed changes that he saw in some landscapes to human-induced climate change, becoming probably the first scientist to seriously appreciate the idea that humans could alter climate, at least on a local or regional scale.)

Throughout the 1800s other scientists developed the idea of ecological succession, the idea that the makeup of species communities develops or matures in stages through time. For example, after a patch of forest is cut down or wrecked by a landslide, the first plants to dominate the cleared area are so-called pioneer species like weedy grasses and annual wildflowers. They are succeeded by larger intermediate species like shrubs, which in turn are shaded out and displaced by a community of tree species, which re-form the forest.

In the early 1900s the American ecologist Frederic Clements began describing communities of plant species as "complex organisms," which could be studied just as individual organisms could be studied. In 1916 Clements published one of the most influential books in the history of ecology, *Plant Succession: An Analysis of the Development of Vegetation*, which detailed his theory of succession. Not only do plant communities progress through well-defined stages of succession, Clements wrote, each stage changing the environment to pave the way for the next, but these stages lead to a final and stable state, what he called a climax community. A specific climax vegetation community is determined by the local climate and is also an indicator of the climate where it occurs. So you can both determine the local climax vegetation by the local climate, and the local climate by the local climax vegetation (this tautology appears to have escaped Clements's notice).

Thus, according to Clements, every natural plant community has a mature state, and it is either in this state or developing toward it following some sort of disturbance. In places that have enough rain and a climate that isn't too extreme, this climax community will be dominated by trees in the form of a closed-canopy forest. By Clements's reasoning, grasses can define climax communities only in places where trees cannot grow—where there isn't enough rain, where temperatures are too extreme (such as in the Arctic or the Sahara Desert), or where soils are too poor or shallow to support trees.

Although other ecologists of the early twentieth century had different views on plant community development and some disputed the idea that communities had stable climax stages, Clements's ideas became popular. The fact that most trees were relatively large and long-lived, and could dominate grasses and shrubs by outcompeting them for soil nutrients and water and blocking their access to sunlight, made their status as climax plants rational to many people. Forests could easily be seen as somehow better or more noble than other vegetation types—as ultimate and most desirable form of plant community.

Clementsians said that grasslands or savannas found in areas with good rain, a moderate climate, and good soil are by definition pioneer or intermediate plant communities, on their way to becoming forests. (Savannas are habitats made up of grasses and trees, but not so many trees that they form a continuous, closed-canopy forest.) In much of Europe, for example, they saw grass-rich areas as those where humans had removed forests in the post-ice-age millennia as they expanded settlements and agriculture across the continent. The existence of fertile, subtropical savannas in Africa was attributed to human burning and cattle grazing.

The fact that some European grasslands contain some very long-lived plant species (including grasses and perennial flowers) and appear to be too species-rich to be such a new, human-created phenomenon was explained away by suggesting that most grass-land species originally evolved in more extreme places (nearer the North Pole, perhaps) and had only recently moved into more temperate areas when humans opened up space for them. Clementsians thought that constant human disturbance like mowing, burning, or intense grazing by domestic animals was necessary to maintain these open habitats as their supposed early or intermediate successional states.

It was only after 1960 that mainstream ecologists began to understand and seek explanations for the fact that vast areas of the planet that have good, deep soil, good rain, and moderate climates have actually been occupied by grass, savanna, and other open plant communities for tens of millions of years, eons before humans or their shaggy, small-brained hominid ancestors roamed Earth. Forests are often *not* the natural climax vegetation in places where trees can theoretically grow. Disturbance by fire and herbivores can beat back trees and turn a dense woodland into a grassy savanna. But if you change the disturbance regime—for example, by reducing the frequency of fire or removing browsers—the ecosystem can flip rapidly back to dense woodland; ecologists call this a system with alternative stable states.

Scientists have now begun to appreciate that grasses and trees have in fact been engaged in constant, complex, multifaceted warfare across very large areas for a very, very long time—at least fifteen million years. They have been fighting to wrest occupation of significant tracts of Earth and its resources away from each other. At many times and in many places, grasses have been able to beat out trees and form long-lasting "disturbance-dependent" communities, even where the climate and soils are ideal for trees. But the longstanding rules of this competition are now being changed by rising $CO_2$ levels.

———

The diversity of life as we know it today would not exist without photosynthesis, the vital process, powered by light (radiant energy) that combines carbon dioxide and water in the presence of an enzyme called RuBisCo[††] to produce complex carbohydrate molecules and oxygen. Photosynthesis typically occurs in special plant cells containing tiny green organelles called chloroplasts. Almost every living creature relies on the carbohydrates produced by photosynthesis to build their bodies and as an energy source to power their bodies; carbohydrates are in this sense the basis of almost all ecosystems.

Photosynthesis likely evolved in primitive, single-celled organisms about 3.4 billion years ago, and its earliest forms probably used not water but other elements and molecules, such as hydrogen, and so did not produce oxygen as a byproduct. (Earth's early atmosphere contained no oxygen, and was likely overwhelmingly

———

[††] RuBisCo stands for Ribulose 1,5-bisphosphate carboxylase-oxygenase. It's one of the most abundant proteins on Earth.

made up of free nitrogen [$N_2$] and carbon dioxide.) About 2.5 billion years ago oxygenic photosynthesis evolved in marine single-celled organisms, but most of the elemental oxygen ($O_2$) they produced was absorbed in ocean water and seabed rock. About 1.85 billion years ago, these below-water sinks became saturated and oxygen began outgassing into the atmosphere, but most of it was absorbed by terrestrial rock. It was only about 850 million years ago that the rock oxygen sink filled up, and atmospheric oxygen levels rose significantly.

It's likely that increasing atmospheric oxygen poisoned many early species into extinction; their physiology simply wasn't evolved to tolerate so much of this "new" element in the air. But it also opened up huge evolutionary space, because oxygen-involved—or aerobic—metabolism can be very much more energy-efficient than anaerobic metabolism. Increasing atmospheric oxygen appears to have helped the evolution of complex, multicellular life forms, which rapidly diversified around the beginning of the Cambrian period, about 560 million years ago. More life in turn meant yet more oxygen, and so our modern atmosphere began to take shape. Increasingly complex types of green plants evolved, including ferns, cycads, and conifers; flowering plants evolved and diversified relatively recently, during the Cretaceous period, 145 to 66 million years ago.

Although some photosynthesis occurs in the green stems of flowering plants, it mostly occurs in their leaves. It requires three inputs: carbon dioxide from the surrounding air, water from the soil, and radiant energy from the sun. The surface of a typical flowering plant leaf is covered in tiny holes called stomata, which are rimmed by so-called guard cells that can open or close them. (Guard cells are a bit like lips around a mouth.) When stomata are open, the outside air—which contains carbon dioxide—can flow into the leaf. Water from inside the plant simultaneously evaporates out; this loss of water from a plant via evaporation through the stomata is called transpiration (and is a very important process in the global climate system).

As water evaporates from the stomata, a water deficit is created in the leaf, which pulls replacement water into the leaf itself via the plant's vascular system—the assembly of pipes roughly analogous to animal blood vessels that transports water and nutrients throughout the plant's body—and creates negative pressure in the plant that pulls water into the roots from the surrounding soil. Liquid water molecules strongly adhere to each other, which is why you can pull water against gravity through a narrow channel like a straw by creating negative pressure at the top end. Put slightly differently, the vascular system forms channels for water to travel from the roots, through the plant, and into the leaf, and the transpiration of water from the leaf creates a suction pressure to pull the water up through the channels.

Thus, during daytime, a leaf can open its stomata to bring carbon dioxide from the air and water from the soil, which combine in the chloroplasts, using energy from the sun to produce carbohydrates and oxygen gas. The carbohydrates are incorporated into the body of the plant, and the oxygen moves out of the stomata into the atmosphere. So long as there is no shortage of water in the soil, a typical green plant can photosynthesize whenever sunlight hits its leaves.

But many plants don't live in environments like rainforests where the soil is constantly moist. They can't leave their stomata open all the time, transpiring continuously, because their roots will pull all the available water from the soil, drying it out, and then their roots will go into water deficit, creating a downward suction pressure. The leaves and roots will pull against each other, and when these opposing suction pressures get too strong, the continuous streams of water inside the vascular system will come apart and the plant will experience hydraulic failure. Bubbles will form inside the vascular cells, and the leaves will no longer be able to pull water upward; it will be like sucking on a straw with a hole in the side, and the plant will die.

Because plants can't photosynthesize without losing water to the air, they must perform a balancing act, opening their stomata for as long as possible to maximize photosynthesis, but not so long that they lose too much water. All things being equal, a plant growing in a hot, arid area grows slower than a plant of the same species in a wet area; the plant in the dry area can't photosynthesize for long each day, and so produces body-building carbohydrates at a slower rate than its conspecific in the wet area.

To cope with vastly different environments, over the span of millions of years flowering plants have evolved three different types of photosynthesis, each with advantages in particular circumstances. The first to evolve was so-called C3 photosynthesis—which is still used by about 85 percent of flowering plant species alive today—followed by types that scientists call C4 and CAM.

The basic differences between C3 and C4 are important because they help explain the structure of many habitats and ecosystems, especially in the temperate and tropical regions of the world. (CAM photosynthesis is mostly used by succulent plants in extremely hot and dry environments and isn't relevant to this chapter, so I'll make no more mention of it.)

Simply put, C3 photosynthetic systems, which are used by trees, work very well at moderate temperatures and when there is abundant soil water, but they can be very inefficient when $CO_2$ levels inside leaves drop to low levels. This drop happens, for example, when the soil is dry and the air temperature is high and a plant must close its stomata to stop losing water to the air. Even though stomata are closed, photosynthetic cells inside the leaves continue to operate if sunlight continues shining on them, using up the $CO_2$ that entered the leaves before the stomata closed.

As the $CO_2$ level inside a leaf falls, RuBisCo, the vital enzyme in photosynthesis, starts causing trouble because it has a design flaw: when there isn't enough $CO_2$ in photosynthetic cells, it starts to "run in reverse" and fix oxygen ($O_2$) instead. This process, called photorespiration, radically reduces carbohydrate production.[‡‡]

In short, by closing its stomata to prevent water loss, a C3 plant effectively shuts its carbohydrate production down. Without carbohydrates, the plant can't grow. C4 plants, which include many warm-season grasses that are common in savannas, have a special, extra "turbo-charger" enzyme which effectively allows them to concentrate carbon and special cells which deliver it efficiently to photosynthetic organelles while keeping oxygen away. C4 plants can reliably flood their photosynthetic systems with carbon, keep the RuBisCo enzyme operating in "the right direction," and prevent damaging photorespiration from occurring. This allows them to fix $CO_2$ with less need to open their stomata; they can photosynthesize efficiently in hot weather and when soils are dry. C4 grasses have a further advantage over C3 trees in that they only have to produce enough carbohydrate to grow leaves and a relatively small root system, whereas C3 trees must grow trunks, branches, and large roots.

These differences between C4 and C3 plants provide a compelling explanation for why grasses and trees have at different times had the upper hand in their long-running war over space, and why the Cheetah is now being pushed out from large areas of Namibia.

---

We often think of plants as immobile, which they generally are in the sense that regular individual plants, once rooted, can't get up

---

[‡‡] High temperatures can also induce photorespiration, meaning that C3 plants photosynthesize optimally at moderate temperatures, below 25 degrees Celsius (75°F).

and move themselves somewhere else in the way that most animals can. But plant species are mobile in the sense that they can progressively occupy new areas (by projecting seeds away from parent plants, for example) or be extirpated from areas that they formerly occupied. Plants, like other organisms, must compete with other species in their ecosystems; plant survival, on both the individual and the species level, can be seen as a battle for space and resources—though a slow-moving one relative to most battles in the animal world.

Grasses may exclusively occupy an area for some time—making it an open prairie—and then get overwhelmed by trees, which will turn the area into closed-canopy forest, and vice versa. Millions of acres of temperate and tropical landmasses are also covered with savannas, which are in-between habitats: they have significant grass cover, but also trees, albeit not enough trees to form closed-canopy forest. They can be seen as battlefields where neither grasses nor trees can yet claim decisive victories, and the war being fought there is mediated by other players, including plant-eating animals and fire.

Scientists have progressively understood more about the dynamics of savannas—how they form and change—over recent decades. They've confirmed that animals ranging from tiny insects to elephants can have a massive impact on vegetation: an insect outbreak can remove the plant species it feeds on from an extensive area, for example, or a large herd of elephants can push over and consume almost all the large trees. Trees or grasses can gain advantage if an influential herbivorous species becomes abundant and targets the opposing side, and quickly lose out if that herbivore disappears.

Fire usually favors grasses, because most grasses are adapted to deal with regular fire. Their leafy tops burn off, and they resprout from the base. (Some perennial grass plants can live for centuries while burning almost every year.) Many savanna tree species need to grow to a certain height—usually about ten feet—to avoid their branches being consumed and killed by deadly flames soaring up from the grasses beneath them. If fires consistently burn across an area at shorter intervals than it takes for trees to grow to a flameproof height, the area will become pure grassland. If fire is kept out for many years, more trees will reach adulthood and shade out grasses, and the area will become a forest.

Models and field observations confirm that stable mixed savannas arise where fires occur on average a little more often than it takes trees to reach fireproof height, but at varying intervals—mostly often enough to eliminate young trees, but occasionally far apart enough in time to allow some trees to reach flameproof height and persist in the landscape.

But herbivores and fire aren't the only factors that determine the winner in Grasses vs. Trees. Rainfall counts, too, because typical savanna trees use normal, old-fashioned C3 photosynthesis, and typical savanna grasses use C4. When rainfall is regular and abundant, trees can keep their stomata open for more of the day, make more carbohydrates, and grow faster. Grasses do relatively better during droughts, because they can photosynthesize for relatively longer each day than trees. In droughts, trees take longer to grow tall enough to escape the fire trap, whereas grasses can continue accumulating burnable mass, thus turning themselves into more effective weapons against their woody rivals. Overall, more—and more consistent—rain favors trees, and drier weather favors grasses.

———

Satellite images and vegetation surveys show that many savannas around the world, from Australia to the Americas to Africa, are rapidly becoming less grassy and more dominated by trees.

In some countries, like Namibia, this is having dramatic economic effects. Namibian ranches now produce less than half the beef they did in the 1950s because millions of acres are now covered in dense thickets of trees where cattle cannot graze. Many ranches have dropped in value, and have begun producing wood and charcoal in an effort to turn a profit on their changed land.

The phenomenon called bush encroachment was first noticed in southern Africa in the late 1800s. By the 1970s it became too big to ignore, especially in Namibia, and agricultural scientists rushed to study it. Many came to the conclusion that it resulted from overgrazing by domestic animals. If you run too many cattle on the land for too long, they will consume the grasses down to the roots and kill them, leaving the soil surface open for tree seeds to establish themselves, they said. Young trees provide negligible fuel compared to grass, so the flames that would normally knock new trees out the landscape came by far less often; trees created their own fire-proof zones. Trees, with their deeper roots, also outcompete grasses for water. The solution to bush encroachment, they said, was for ranchers to cut down some trees, keep cattle off the land long enough for grass to reestablish itself, and then burn regularly to favor grass.

Many ranchers did this, but often to no avail. The trees kept coming, becoming denser and denser over more and more land every year. Ecologists in neighboring South Africa, studying historic photos of landscapes in their national parks, saw that trees were taking over there too, in natural savannas that were not overgrazed by cattle and that had experienced no change in fire frequency or rainfall. North and South American and Australian researchers saw "woodification" in many of their savannas. Scientists noted that trees were even spreading in some savannas where average air temperatures have been rising, even though hotter temperatures should theoretically favor C4 grasses, not C3 trees.

After considering the influence that atmospheric $CO_2$ has on different plants' growth, many scientists now think that the triumph of the trees is due to changes in the atmosphere that are eroding a battlefield advantage that grasses enjoyed for thousands of years. They have observed that C4 grasses' special "turbo charger" adaptations not only allow them to photosynthesize efficiently when their stomata are closed in hot, dry weather, *but also when atmospheric $CO_2$ levels are low.* Put another way, the atmospheric $CO_2$ level doesn't matter that much to C4 plants; even if there's very little $CO_2$ in the air, they effectively concentrate carbon and feed a high-quality supply to their photosynthetic systems.

But atmospheric $CO_2$ really matters to "old-fashioned" C3 trees. They can't concentrate carbon and keep oxygen away from RuBisCo like C4 plants can, so they theoretically can't photosynthesize at high rates in a low-$CO_2$ atmosphere, even if everything else is in their favor; even if temperatures are moderate and they have access to abundant water. (Atmospheric $CO_2$ levels had been relatively low since before the Pleistocene epoch, which began about 2.6 million years ago, but now, of course, they're rising fast.)

Scientists have experimentally grown common African savanna tree species in special chambers containing different levels of $CO_2$ in the air, ranging from 180 parts per million, the level at the Last Glacial Maximum (about twenty thousand years ago) to 400 ppm (2015's level) to 1,000 ppm, a level predicted under extreme scenarios of continued fossil fuel burning. The results are stunning: A recent study found that increasing atmospheric $CO_2$ from 180 ppm to 400 ppm caused seedlings of a common southern African savanna tree, Sweet Thorn (*Acacia karroo*) to grow 53 percent faster. When air $CO_2$ was raised from 400 ppm to 1,000 ppm, seedling growth increased a *further 230 percent.* These experimental studies have also shown that tree seedlings grown in higher-$CO_2$ air recover from fire and grazing injuries faster and more reliably. This is strong evidence that many young savanna trees have been growing well below their potential for thousands of years and are now being fertilized by rapidly increasing atmospheric $CO_2$. The same type of experiments with common African C4 grass species show little change in growth rates at different levels of atmospheric $CO_2$.

Imagine a generic southern African savanna that burns once every four years on average: a few decades ago, the trees in this savanna grew at a rate that would get them to a fire-safe size in five years. If, every now and then, there was a gap between fires of longer than five years, a few trees would reach fire-safe maturity and persist in the landscape. But with rising $CO_2$ levels, trees are growing far faster than before: let's say they can now reach fire-safe sizes in just three years, a shorter period than the average interval between fires. This gives many more trees the chance to escape the fire trap, and unless fires consistently come significantly more frequently than in the past, trees will shade out grasses and take over in just a few years.

It bears restating that savannas are complex, and changes in many factors including rainfall, fire, and grazing can push them in a more-grassy or more-woody direction. But increasing atmospheric $CO_2$ has put a large and swelling thumb on the scale for trees.

---

Fossil beds across southern and eastern Africa are filled with long and detailed stories of open ecosystems. Deep down you'll find layers densely strewn with fossilized C4 grass pollen and charcoal, proving that grasses and the fires that promoted them have been here for tens of millions of years. They've long been evolving into an ever-radiating variety of species, their rich leaves supporting vast herds of grazing animals and the predators and scavengers that feed on them.

Fossil bones in the rock show animals evolving for landscapes with ever-more light and space. As open habitats persist and become larger over time, we see a range of gazelles and similar species like Springbok and Impala appear. Antelope legs become longer, their bodies lighter, and their muscles develop to allow them to run faster, because raw speed matters more than camouflage or maneuverability when escaping predators in the open. A rich variety of grassland birds evolves, with long legs for running and higher-pitched voiceboxes to carry their songs farther in treeless lands. Cheetah ancestors become faster, too.

The ancestors of modern-day ostriches lose their ability to fly in favor of evolving extraordinarily strong, long legs. But one grass-loving species, the Southern White Rhinoceros, barely changes at all: skulls from three million years ago are identical to those in today's living rhinos, likely because its large size allows it to shape the ecosystem to its needs. It persistently consumes tall-grass species until they disappear and are replaced by succulent, protein-rich shortgrasses, which it then crops down to lawn length; Southern White Rhino territories are often centered on these grazing lawns, which they maintain for their whole lives.

As more mammals have evolved specializations to live in grasslands, so too have scavengers. More herbivores and more predators mean more carcasses to feed on. There has been so much leftover meat on grasslands and savannas for so long that a whole group of birds of prey, the large African vultures, have evolved a suite of specializations to succeed here. Unlike the New World vultures of North and South America, which fly relatively low and use their excellent sense of smell to find rotting carcasses in thickly wooded areas, African vultures use their eyes. Their long, broad wings are optimized for hyperefficient gliding on rising hot air, not maneuverability or speed. As the day's air temperature rises, they slowly circle thousands of feet up into the air, scanning massive acreages of land with extremely high-resolution vision; they can fly hundreds of miles in a day in search of new carcasses. When a vulture sees a carcass, it pulls its wings in and drops as fast as possible, other vultures nearby doing the same. Within minutes a carcass can have hundreds of vultures on it, hissing and fighting, tearing into flesh, gulping it down as fast as they can.

---

Five hundred years ago, just as Europeans began colonizing, Cheetah were found across much of southern Africa. Only a tiny percentage of the subcontinent was covered by true closed-canopy forest. Much of the east was filled with wooded savanna of various kinds, creating fair habitat for Cheetah and other predators, and most of the west was drier, more open country—ideal for the cats—each part with its own geology and unique species: the Karoo of modern South Africa, with an extraordinary diversity of succulent plants; the red-sand Kalahari semidesert region, occupying parts of today's South Africa, Botswana, and Namibia; the various deserts and semideserts of Namibia, from the bone-dry Namib to the slightly moister grasslands and savannas farther north.

All these western regions were rich in grasses, and despite their aridity were home to vast numbers of plains game like Hartebeest, Oryx, and three or four types of zebra. The most numerous species across this vast area was the Springbok, a gazelle-like medium-sized antelope with elegant lyre-shaped horns, a white face, a pale golden-brown neck and back, a striking broad dark-chocolate-brown stripe along its side, and a white belly. Its long, thin legs, tipped with sharp black hooves, can carry it at great speed across the plains—it's often called the fastest antelope in Africa—and its name alludes to its exceptional jumping abilities; an adult Springbok can clear about seven feet.

Millions of Springbok lived across the western half of southern Africa five hundred years ago, and helped to support thousands of Cheetah. (Although other predators target Springbok, only the Cheetah is fast enough to reliably take down adult Springbok in open country.) Springbok were nomadic; after widespread good rains, they would disperse across the landscape to feed on new growth. Once they'd grazed a region almost to its limit, they would group into progressively larger masses, and then, as one, with no obvious leaders, begin trekking for hundreds of miles until they reached good veld again. Predators and scavengers followed the herds as they moved.

European settlers recorded Springbok herds of well over a million animals apiece in the 1700s and 1800s. They took days to pass by single points, shoulder to shoulder, nose to tail. Settlers sometimes killed Springbok with billy clubs as the herds passed through settlements; the animals were so crowded together that they couldn't run away.

Springbok masses shaped grasslands wherever they moved. Their hooves broke up the soil, aerating it and allowing the little rain to infiltrate before evaporating. Their droppings fertilized the veld. They could act like fire, consuming and trampling vegetation at speed, leaving perfect short-stubble fields behind for ground-nesting birds to use, and triggering grasses to send up fresh green shoots for other herbivores, who followed a few weeks behind. They were the ecosystem wealth generators of this part of Earth, but their influence was not to last.

Increasing numbers of white settlers saw Springbok as competition for their cattle and sheep. They shot them—and any other wild animals they could find—without restraint. At first it seemed as if there would be no end to the herds, but as the 1800s progressed the hunting parties returned with fewer and fewer animals of smaller and smaller species. By the late 1800s some could find only hares to kill. The last Springbok trek took place in 1897; after that there were too few animals, too dispersed across the landscape to trigger their massing behavior.

The settlers put up thousands of miles of fences to enclose their cattle and sheep across South Africa and southern Namibia, further depressing plains game numbers. The few Cheetah that survived here turned to eating domestic stock, and they were soon shot by ranchers in return. Their day-hunting, open-country habits made them easy to find.

By the 1950s there were almost no Cheetah left in South Africa or southern Namibia. Some survived in national parks and a few hundred roamed the wild center of the Kalahari—mostly in Botswana—with more in northern Namibia, where ranches were vast and lightly controlled, and where scattered groups of trees gave Cheetah places to hide from the guns.

In the 1970s conservationists began to recognize how rare Cheetah had become—they had been almost completely wiped out in Asia and been lost from many African countries—and began working with northern Namibian ranchers to conserve the animals on their land. Over many years, they changed public attitudes toward the cats, and together with ranchers developed ways of keeping cattle safe from their claws and teeth. Namibian Cheetah populations stabilized, and by the 1990s it had the largest remaining Cheetah population of any country, some 1,500 animals.[§§]

Especially large numbers roamed the Otjozondjupa area in the 1980s, including the great-great-grandmother of the Cheetah who

---

[§§] Namibia still has about 1,500 Cheetahs today. The total global population is about 7,000.

opened this chapter. That matriarch fed mostly on Springbok and Hartebeest, fast-running, grass-eating midsize antelope species. She bred well and raised many young, one of which was "our" Cheetah's great-grandmother, who took over the territory when her parents passed on.

By the 1990s trees began to crowd into Otjozondjupa. Ostrich, Hartebeest, and Springbok numbers began to drop. New species of browsers, able to exploit trees, moved in: massive, statuesque Kudu, the males with towering spiral horns, adults far too large for Cheetah to tackle, and small Gray Duiker, whose low-slung bodies allowed them to duck easily into thickets.

By the time "our" Cheetah was born, around 2010, many midsize antelope were gone, pushed out by trees. She could find enough game to live on, but not thrive—her offspring moved far away, to places where some openness remained.

Large vulture species declined along with the Cheetah. When an animal dies in a thicket, it's much harder for vultures to see it. Also, by the end of a good meal, vultures have usually consumed so much meat that they can barely fly. Taking to the air again requires them to run, flapping hard, sometimes for over a hundred yards before they can reach enough speed to counteract gravity. This is not a problem in open, grassy landscapes—just about everywhere is a runway—but when there are too many trees, too close together, full vultures simply can't take off. In the days that it takes to digest their meals they're grounded and vulnerable to attacks by predators.

So the giant birds avoid searching over and landing in densely wooded areas. Every acre of open savanna that transforms into thicket is another acre denied to them, and with large African vulture species already in decline and endangered by poisoning and collisions with power lines, they need every carcass they can get.

Today, with the Cheetah gone, the days are less dangerous for Gray Duikers, Steenbok, and the few young Kudu that still live in the thickets of Otjozondjupa. Leopards hunt mostly at night, so the small antelope can drop their guard a little in the daytime, moving farther out into the shrinking open areas that remain to graze on the grass that holds on there.

You can imagine the spirit of a Cheetah who died generations back returning now to visit his old places, to trace the same daytime paths he used to walk across Otjozondjupa. Instead of striding across dull-yellow grassveld, he is ghosting through packed, thorn-covered brown branches. He can see for just a few yards, and nothing like a horizon of any kind. A low canopy hangs overhead, casting a random matrix of dense, overlaid line-shadows broken up with random tiny spots of sharp light. All around are hundreds or thousands of rigid woody stems, barring direct passage to any creature larger than a hare.

Even the soundscape of the new habitat is claustrophobic. Rattling Cisticolas—tiny gray-brown birds that thrive in thorny, woody savannas and have colonized in large numbers—sing throughout the day, even in the midday heat, which is *hot*. Perched up near the tips of thorn-covered twigs, they assertively jerk open their tiny bills and, with vibrating tongues and shaking bodies, tirelessly churn out their loud territorial song, three strident notes followed by a lower, sharp, rattling *churrr*:

*Chew! Chew! Chew! Churrr.*
*Chew! Chew! Chew! Churrr.*
*Chew! Chew! Chew! Churrr.*

They are everywhere, every hundred yards or so, in every direction. As he moves away from one singing bird, he begins to hear its neighbors. Sometimes their tone, timbre, or tempo varies—*Cheeoo! Cheeoo! Cheeoo! Churrr,* or *Tew! Tew! Tew! Chrrp,* or *Teeuw! Teeuw! Teeuw! Chrrrrrup!*—but it's almost always three notes, then a rattle. Three notes, then a rattle. Three notes, then a rattle. Three notes, then a rattle, over and over. The thicket covers the land for miles in every direction, and there is no escaping this noise.

7

# Sea Change

Australia's Great Barrier Reef complex runs alongside more than fourteen hundred miles of the northeast coast of that continent, from the northern tip of the Cape York Peninsula southward to Elliot Island. It's Earth's largest structure made by living creatures, formed of roughly three thousand individual reefs and nine hundred islands that occupy an area the size of Germany. Today's reef complex is relatively young—it began forming about nine thousand years ago, as the last ice age waned and the oceans began to slowly rise and warm, creating good conditions for trillions of tiny coral polyps to settle here and make homes for themselves, in turn building habitat for a vibrant, otherworldly array of thousands of other species.

Over sixteen hundred species of fish live on the Great Barrier Reef, along with over three thousand types of mollusk. Three-quarters of the world's coral species are found here, including over four hundred types of stony corals—those that build and make up the hard body of the reef. Different species of stony corals grow in different forms, including massive corals—dense, round, large, and slow-growing—which make the basic, solid foundations of reefs; tabulate corals, which are large and flat, like tables; laminar corals, like shelves; foliose, like leaves; encrusting, which grow over surfaces; digitate, like forests of small fingers; corymbose, whose dense, irregular branches create excellent hiding places for small fish; and branching corals, with long, splitting, treelike branches, reaching up toward the light. There are also solitary corals; isolated forms, usually around the edges of reefs.

---

Faceless but beautifully formed, brainless but coordinated, motionless but alive. The identity of corals confounded philosophers and scientists for thousands of years. Were they plants or animals, or a combination of plants and animals—or something else entirely, perhaps some strange form of rock? Pliny the Elder said that corals were neither plant nor animal, but possessed a "third nature." The Babylonian Talmud classed them as trees. Most westerners considered them to be some sort of plant until the late 1700s, when the famed British scientist William Herschel examined them under a microscope and declared their cell membranes to be positively animal in nature.

Today scientists classify corals as animals in the phylum Cnidaria, the high-level group of aquatic invertebrates that includes jellyfish and sea anemones. Most corals live as colonies of tightly packed, genetically identical polyps, each polyp being a saclike animal typically a few millimeters in diameter and up to a few centimeters long. Each polyp has a tiny mouth opening surrounded by minute tentacles. There are two major groups of corals; soft corals, which have no hard exoskeleton and whose polyps have eightfold symmetry, and hard or stony corals, which have rigid exoskeletons made of calcium carbonate and whose polyps have sixfold symmetry. The latter are the primary reef builders, although many soft corals live on reefs too, and are important constituents of their ecosystems.

To build their tubelike skeletons, stony corals draw carbonate ($CO_3^{2-}$) and calcium ($Ca^{2+}$) ions from seawater to produce aragonite, a form of calcium carbonate. (Many other sea creatures, like mussels and clams, also make aragonite shells this way.) Different species of coral create different colonial structures, each typical of its species. As the polyps age, they accrete more and more aragonite and their skeletons become longer. When they die, younger polyps build their skeletons around and on top of their deceased forebears' refuges. Healthy reefs can grow by several centimeters every year.

Timing is everything for coral reproduction, and most species spawn just once a year. Mature corals develop eggs and sperm within the polyps and then, just after sunset at a particular phase of the moon, usually in the late summer, release them simultaneously. For a few hours the moonlit water is filled with swirling galaxies of coral gametes, floating up to the surface, male and female swirling together. They will join to make tiny larvae that grow a little and then settle out, attaching themselves to a substrate—which can be bare rock or part of a living reef—and begin to build their skeletons.

Although they're classified as animals, many coral species live as more-than-animals, in tight symbiotic associations with single-celled algae—that is, minute plants—called zooxanthellae. The algae grow in special tissues within the polyps, consuming the polyps' carbon dioxide, phosphate, and nitrogenous waste compounds. Via photosynthesis, they turn these waste products into oxygen, sugars, and carbohydrates which the polyps use to live and grow. This is why almost all stony corals grow in shallow water, up toward the sunlight—without it, their green symbionts would die. Although coral polyps can and do feed on small animals like plankton or even tiny fish (which they pull into their mouths with their tentacles), most species gain the vast majority of their nutrients from their partnerships with zooxanthellae.

But sometimes corals lose their zooxanthellae and are forced to go it alone. Stress can cause the polyps to expel the microscopic algae cells en masse. Since most corals get their color from their zooxanthellae, this tends to turn corals pale and white, hence "coral bleaching." Pollution and disease can trigger this expulsion, but it's most often caused by an above-normal increase in water temperature—an ocean heat wave.

There are several mechanisms by which symbionts are expelled, not all of which are well understood. The current scientific consensus is that expulsion is triggered by oxidative stress. As water temperature rises, so does the concentration of thermal energy within the polyp, which sends the symbiotic algal cells' rate of photosynthesis into overdrive. The symbionts start to produce more oxygen than the polyps can absorb, and the excess oxygen ions combine with other elements to form toxic molecules. This irritates the polyps, and they break open the cells in which the symbionts are hosted and push them out into the water. Once about 70 percent of the algal cells are expelled, coral bleaching becomes obvious to the human eye.

---

Earth has had oceans since about 150 million years after it was formed about 4.5 billion years ago. As soon as the new planet's temperature dropped low enough, water began condensing out of the early atmosphere and covering the surface. The first microscopic life appeared in the oceans in the Archean period, about 4 billion years ago, only making it onto early land 300 million years later. Many of the evolutionary innovations that gave rise to life as we know it today arose in the oceans; the first vertebrates lived here, long before they evolved terrestrial forms.

Today the oceans occupy about 70 percent of the planet's surface, but they have absorbed more than 90 percent of the excess energy that human carbon emissions have caused to be retained in the biosphere. They're also a hugely important carbon sink, although as they absorb more carbon gases, their water chemistry changes, with serious effects on marine creatures.

Water has a far higher thermal energy capacity than air, which is to say that it preferentially absorbs energy and can acquire more thermal energy without its temperature rising nearly as much as air; this is why the ocean has absorbed relatively more thermal energy than air has. When thermal energy enters the oceans, it tends to stay for decades, circulating ever deeper, holding on to it far longer than air. Without the ocean, Earth's atmosphere would now be several degrees warmer.

The ocean also absorbs atmospheric carbon dioxide, which reacts at the water surface to form carbonic acid in the water. Carbonic acid then splits into hydrogen ions and bicarbonate ions, and the latter get used by ocean creatures to build their skeletons and shells. All ocean creatures, from the tiniest plankton to the largest whales, eventually die and fall through the water column, all the way to the seabed, unless they're intercepted by a scavenger on the way down. Their remains become buried under subsequent layers of dead creatures; they become part of the sediment of the ocean floor, and their carbon is locked away from the water and air, effectively forever. (Before industrial whaling devastated whale populations worldwide, whale skeletons likely sequestered many tons of carbon every year.)

But if atmospheric carbon dioxide levels rise quickly, this can seriously harm ocean life and ultimately curtail the oceans' ability to absorb carbon. As stated just before, one of the products of carbonic acid when it splits in ocean water is hydrogen. The more hydrogen ions water contains, the more acidic it becomes, which reduces the availability of carbonate ions in the water. Hard-shelled animals, including corals, that need to pull carbonate ions from the water to build their shells find this increasingly difficult to do.

Because about one-third of the $CO_2$ emitted by humans is absorbed by the oceans, they have already become measurably more acidic since the Industrial Revolution, dropping about one-tenth of a pH point since that time. Young shellfish in some coastal areas, like oysters in the Pacific Northwest of North America, are already growing thinner shells than before. Past mass extinction events have seen acid spikes in the oceans and a resulting sharp reduction in hard-shelled organisms, including many calcium-rich plankton species that are important food for young fish, causing further extinctions and ecological disruption. Oceanic pH is projected to drop by another two-tenths of a point by 2100, faster than in over three hundred million years.

Changes in ocean pH don't just affect hard-shelled animals; they influence enormous numbers of other chemical reactions in the ocean. Some reactions speed up, and others grind to a halt. Nutrient cycles—like the vital nitrogen cycle—are disrupted and changed. The network of knock-on effects is so complex that they are virtually impossible to predict, but they are profound. These disruptions can so severely affect ocean ecosystems that they become far less effective in their current role as long-term carbon sinks.

South Africa is bracketed by two large and significantly different ocean currents. The warm Agulhas Current, which flows south from the tropics and along the country's eastern and southern coasts, and the cold Benguela Current, which flows north from the sub-Antarctic and along the west coast.

Like many warm ocean currents, the Agulhas carries relatively few nutrients, and produces relatively little biomass, but as the wind passes over the cold Benguela along the west coast of South Africa, it triggers strong upwellings of deep ocean water, which bring up nutrient-rich sediment, feeding vast shoals of small pelagic fish like anchovies and sardines. The Benguela supports a bountiful marine ecosystem which includes a unique bird, the African Penguin, a medium-sized black-and-white species that has lived around the southwestern corner of Africa for a very long time.

Being flightless, the African Penguin is vulnerable to land-based predators such as jackals and Caracal Lynx, so it needs to nest in places that these species can't reach. Luckily, there are twenty-one small islands along the west coast of South Africa and neighboring Namibia, ideally placed just offshore, in the middle of the fish-rich upwelling zone. The millions of other fish-eating birds—four species of cormorants, Cape Gannets, and the like—that have also nested on these islands since time immemorial have produced vast quantities of guano, forming a deep soil-like layer on the islands for penguins to tunnel into for nesting burrows. A yard or so into the substrate is a shady, temperature-stable place to lay two eggs, which is very convenient because summer air temperatures regularly go over 35 degrees Celsius (95°F) here.

Male and female penguins trade twenty-four-hour shifts incubating the eggs. As one goes out to sea with a foraging flock of tens or hundreds others—African Penguins are expert collaborative hunters—the other waits patiently on the eggs in the near darkness, making sure they remain neither too hot nor too cold. When its mate returns from the ocean, the pair greet each other with harsh, excited, donkey-like brays before swapping roles.

Hatchling penguins are fluffy chocolate-brown and voracious. As soon as they're large enough to thermoregulate alone, both parents leave the nest to hunt together, swimming up to fifteen miles away from the colony to fill themselves with fish. They return at least once every twenty-four hours, waddling onto the shore to vomit up their silvery catch into the gullets of their eager, whining babies. After a few minutes' rest on the beach, they head out to sea again, "flying" underwater with grace and speed.

About a million pairs of African Penguins lived around southwestern Africa a century ago. But then fertilizer companies found their islands, and began clearing them of guano to be sold at handsome prices to farmers around the world. There was no place left on many islands to dig good nesting burrows, so the penguins began laying their eggs out on the surface, where they were vulnerable to overheating and predation by large, black-backed Cape Gulls. Penguin eggs were harvested and sold as delicacies. Then, after World War II, the industrial fishing fleets moved in, scooping up millions of tons of anchovies and sardines to grind up into fishmeal or stuff into cans.

At first it seemed as if the Benguela's fish stocks would never stop replenishing themselves. The never-ending supply of nutrients sustained extraordinary fish growth rates, and each female anchovy or sardine could produce thousands of eggs. Fishing wasn't well controlled. But African Penguin numbers were dropping fast. By 1970 only about three hundred thousand pairs remained, and in the 1980s the stocks of small pelagic fish off the Namibian coast crashed to less than 5 percent of historical levels.

Penguin babies there began starving, and by 1990 the species' total population had fallen below one hundred thousand pairs.

Fish populations survived in the southern part of the African Penguin breeding range because South African fisheries remained fairly well regulated as the millennium dawned. But despite good management of key fishing areas around South Africa's west coast, the small fish that penguins feed on began to decline soon after, followed by the penguins. By 2010 only forty thousand pairs of birds remained, many around the city of Cape Town.

Biologists have found that most of the small fish remaining are now living around the southern coast of South Africa; they've moved from the west coast. They're not sure why, but climate change is causing major ocean currents in this region to become far less stable and predictable. Small fish are more sensitive to environmental change, and it's likely that they're moving in response to a climate driver that scientists can't yet understand.

Unfortunately for penguins, there are no breeding islands along an almost six-hundred-mile-long stretch of the southernmost African coast. Although nonbreeding birds can and do swim here to feed, it takes them almost a week of fast swimming from the nearest breeding islands to reach the heart of the new good fishing zone. There is no way that breeding birds on the west coast and around Cape Town, limited to a small radius around their nests, can access these new fish populations.

In late 2022 the total African Penguin population had declined to fewer than twenty thousand pairs and was declining at a rate of 5 percent per year. The species is projected to become functionally extinct by 2035, meaning that there will be so few of them that they can't breed successfully or maintain their role in the ecosystem.

Increasingly desperate conservationists are trying to prevent this doomsday scenario. They're building handmade concrete shelters on breeding islands, double-layered for insulation, to shelter penguin eggs from rising temperatures. They've also constructed an "artificial island" on the southern coast, where the last small pelagic fish populations now live; this is not a real island, but a small rocky headland that juts three hundred yards out into the ocean.

In 2018 they built a six-foot-high fence to cut this headland off from the mainland, preventing the local Cape Leopards and other predators from reaching it, and placed dozens of pairs of life-size concrete model penguins on seaside rocks. Young penguins rescued from failing nests in faraway colonies and raised by hand were released here. Powerful speakers blared recorded brays out to sea day and night to attract birds to the area. For years this effort appeared to be in vain—although a couple of penguins out on extended fishing trips rested here overnight, there was no sign of permanent habitation or breeding until August 2022, when an observant conservationist noticed two fluffy brown young poke their heads out from underneath a resting adult penguin. Perhaps, just perhaps, a new colony will start to form where their best food fish now swim.

---

The ocean is relatively easy for its creatures to move across compared to land. There are no mountain or river barriers. If water conditions are suitable, fish can swim pretty much wherever they want. As oceans have warmed in recent years, huge fish migrations have changed course. Millions of herring are now spawning over five hundred miles farther north in the North Atlantic than they did in the year 2000. Arctic Cod, which used to have the northern polar regions all to themselves, are now competing with other species of cod that have moved north. Since 2007, vast shoals of Mackerel have moved away from the European continental edge and are now living off the coasts of Iceland and Greenland. Red

Mullet and Anchovy are now common in the North Sea off Britain; just a few years ago they were found only off the Iberian Peninsula. The famous lobster fisheries of New England are in rapid decline as the animals move north from their familiar habitats off the coast of Maine into Canada. Combined with overfishing, this massive and ongoing change in many fish populations is disrupting and breaking fishing industries and marine food chains around the world.

Since 2019, tens of thousands of dead seabirds have been washing up on the west coast of Alaska during summer. They're not being killed by disease or new predators; they're starving, as their emaciated breast muscles show. The fish they feed on used to migrate into the Bering Strait at that time of year but are now heading farther north, into the rapidly warming and increasingly ice-free Arctic Ocean.

The warmest oceans around the equator are fast losing their fish, too, as their temperature rises and their species move to higher latitudes to stay within a livable temperature range. Fish that evolved around the equator have always lived close to the upper limit of fish's intrinsic thermal tolerance; there are no hotter ocean waters anywhere nor fish that can live at higher temperatures. So as equatorial waters warm, there are no species to replace those moving away. A 2020 study showed that the average marine fish species is moving about fifty miles away from the equator every decade—a huge shift, causing massive and largely unknown knock-on effects throughout the entire biosphere.

---

Early December 2022: as the sun goes down over tiny Raine Island, a small coral cay in the northern Great Barrier Reef, a female Green Turtle waits in the darkening water just beyond the reef. She was born from a soft egg buried on the beach a hundred yards from here a quarter century ago, and it's her first time back since she hatched, pulled her way up through the sand, and scrambled toward the waves. She mated in the water with a waiting male an hour ago. As soon as it is fully dark she will swim to the beach, crawl up to the first dune, laboriously dig a hole (her flippers are designed for swimming, not digging), lay about a hundred eggs in it, cover them up, and then crawl back to sea.

She won't be alone. Tonight another sixty thousand or so female Green Turtles will nest on this minute piece of land, as they have for thousands of years. They'll hang around the island for about two months, returning to lay new batches of eggs every two weeks or so during that time. By the end of the season they'll each have put between seven hundred and a thousand eggs in the sand.

Green Turtle young take about eight weeks to mature in the eggs, then they all hatch around the same time. Small, tasty, and vulnerable, many will be picked off by seabirds as they scuttle across the sand toward the water. They'll encounter more predators like large tuna and sharks in the ocean. It takes about twenty-five years for a Green Turtle to reach sexual maturity, but only about one in a thousand hatchlings will survive that long. Green Turtles can live for a century, so the average female adult will get multiple chances to replace herself before she dies.

This breeding season on Raine Island appears no different than any that have come before—tens of thousands of turtles laying millions of eggs at the same time of year as they normally do, hatchlings emerging in masses as they normally do, adults departing for deeper waters around the end of March. But careful observers have noted one extraordinary change from just twenty years ago. Instead of the sex ratio of hatchlings being about 1:1 male to female, 999 out of 1,000 hatchlings this year are female. Raine Island's nests will produce almost exclusively female turtles, as they have since 2017.

Green Turtle broods on the southernmost islands of the Great Barrier Reef (farther from the equator) will also skew female this year, but not so thoroughly—about 70 percent of the hatchlings will be girls, but that's up from about 30 percent just twenty years ago. Australia's Green Turtle population is feminizing rapidly, because of the strange way sex is determined in this species and a subtle increase in thermal energy in beach sand.

Most turtle species (both freshwater and saltwater), all crocodile species, and some lizards determine the sex of their young in the egg by exposing the embryo to large amounts of either testosterone or estrogen during a critical phase of development. (Most other reptiles are like humans, determining sex via genetics.) Scientists don't completely understand the physiological mechanisms for regulating the amounts of these hormones in the egg, but it involves an enzyme called aromatase, which helps to convert testosterone to estrogen.

Aromatase becomes more active at higher temperatures, converting most of the testosterone in the eggs to estrogen. Green Turtle embryos exposed to large amounts of estrogen become female. Thus eggs incubated at high temperatures become male, cool nests produce females, and those incubated at intermediate temperatures produce a mixture of sexes. The temperature difference between a nest producing only males and a nest producing only females is about two degrees Celsius.

The pro-female bias in hatchling Green Turtles is already feeding into the adult population: at least 60 percent of Great Barrier Reef adults are now female. Because one male Green Turtle can mate with several females, this isn't likely to hurt Green Turtle reproduction rates in the short term—reproduction rates may even go up. But as the older adults die out, the population will trend toward being 100 percent female, unfertilized eggs will become more common, and unless Green Turtles find somewhere cooler to lay their eggs—unlikely, because females have a strong drive to return to the beaches where they hatched—their population will crash.

Great Barrier Reef turtles aren't alone in becoming female: a recent study found the same pattern in a large Green Turtle population in Equatorial Guinea (on the west coast of Africa), and is happening elsewhere too.

And what of all the other species whose young become male or female depending on egg incubation temperature—the freshwater turtles, lizards, and crocodiles? Few are currently being studied in the wild, so we don't know how rapidly their sex ratios are being pushed out of whack by climate breakdown.

—

The shallow waters along the west coast of North America are populated with Giant Kelp, a species that at first sight looks like a massive underwater plant. It has what appear to be roots anchoring it to the seabed and a smooth "stem" (correctly called a stipe), several inches in diameter, that can grow up to sixty yards long. It's topped by what appear to be floppy leaves that float just below the ocean surface, rippling in the current. But although this extraordinary creature photosynthesizes like a green plant, it is not one; it's a type of brown algae—in fact, the largest-bodied algae on the planet.

Giant Kelp grows in cool, nutrient-rich coastal waters around the world, from the west coast of North America to South Africa, the southern parts of South America, and the southeast coast of Australia. It forms underwater "forests," filling the role of trees. Moving shafts of light play down from above through its kelp fronds, illuminating seals and fish that fly between its stipes like aquatic birds and, below, an ocean floor covered with all sorts of strange invertebrates.

Giant Kelp can grow up to two feet per day. Although few creatures eat it—there are essentially no fish adapted to bite into its stems or leaves—it's an important base for numerous food chains. Because they grow so close together, the tips of Giant Kelp fronds are constantly being pushed into each other by the waves. As they rub together, wear away, and break up, they release small particles of algae into the water, where they are consumed by plankton, which in turn feed small fish, which feed larger fish, and so on.

The shady, swaying kelp is great habitat for predatory fish, including small sharks, which use its cover to stalk prey. But one of the most fearsome predators of the North American kelp forests is the Sunflower Sea Star, which sightlessly roams the shallows in search of slow-moving invertebrates. Adult Sunflower Sea Stars are over three feet across—larger than a trash can lid—and have sixteen to twenty-four yellow-orange limbs, fringed in purple, which they use to grasp and kill their prey. Among their favorite foods are urchins, whose long protective spikes can't stop the all-enveloping sea star from killing them. Sunflower Sea Stars aren't too fussy about the specifics of their habitats, and are found from Baja California northward to Alaska, where their populations ran well into the billions until 2013, when they—and other species of sea stars—suddenly became sick.

That summer, divers noticed huge numbers of sea stars losing their color and then, within hours or days, literally melting away. Random parts of their bodies would become white and then turn to mush, which dissipated in the water. Because sea stars have no brains or central nervous systems, individual arms or bits of arms would walk away on their own as the rest of their bodies disintegrated. Within weeks entire bays would lose all their sea stars, and it was happening across the entire range of the Sunflower Sea Star, but especially in the south.

During winter, a few juvenile Sunflower Sea Stars reappeared in their familiar haunts, but when summer came around again, they died in the same grotesque fashion as in the previous years. This pattern has repeated itself every year since, with ever-diminishing numbers of young appearing in the winter. In 2022, surveys found only 1 percent of the former population of Sunflower Sea Stars along the coast between Mexico and British Columbia, and only 13 percent of the former population along the Alaskan coast. A major player in the ecosystem is effectively extinct, and this has changed the behavior of their favorite prey, sea urchins, which in turn has totally transformed the undersea ecosystem.

Before the sea stars vanished, sea urchins were extremely wary. They spent most of the time wedged in rock crevices with their spines sticking out, where it was difficult for sea stars to grab and kill them. Urchins eat various types of seaweed—including kelp—and the underwater "landscape of fear" created by sea stars meant they consumed only the seaweed nearest safe refuges. Within months of the sea stars' disappearance, urchins became bold, moving far from their hideouts and consuming all the seaweed they could find. They rapidly created "urchin barrens," large areas where the undersea rocks are scraped clean of all seaweed, vast ocean deserts populated largely by their spiky selves. They have even consumed 95 percent of the Giant Kelp off California's coast, removing habitat for fish and dozens of other species. The only areas where kelp survives have dense populations of Sea Otters, which have taken over from sea stars as the largest predator of urchins. Sea Otters, which just thirty years ago were virtually extinct in California, have been rescued by concerted conservation action, and are now gaining new worth.

What caused the sea stars to disappear? Again, it took them years, but in 2020 scientists found that Sea Star Wasting Syndrome—what the melting-away has come to be called—is the result of

Above: Eucalyptus woodland in Australia's "dry world," Morton National Park, New South Wales, Australia, February 2020

Below: Gondwana rainforest in Australia's "wet world," New England National Park, New South Wales, Australia, February 2020

Superb Lyrebird

Superb Fairywren

All photos © ADAM WELZ

Former pine forest near South Lake Tahoe, California, 2019. This area burned in 2005 and has changed to low, grassy scrubland.

Ponderosa Pine, Taylor Creek Visitor Center, California, 2019

Fire ash/mud mixture in Mannus Creek, New South Wales, Australia, February 2020

Mannus Creek in the Bogandyera Nature Reserve shortly after fire, New South Wales, Australia, February 2020

Bwabwata National Park, northern Namibia
*Above:* Before bush encroachment, November 1999
© LYNN HALSTEAD & GARTH OWEN-SMITH/IRDNC
*Below:* After, September 2019
© CONOR EASTMENT

*Above:* A healthy section of the Great Barrier Reef, Australia, prior to coral bleaching, January 2005 © MIA HOOGENBOOM
*Below:* Bleached coral at Lizard Island, Great Barrier Reef, Australia, 2016 © THE OCEAN AGENCY

Healthy Giant Kelp forest in False Bay, Cape Town, South Africa, 2022 © RUBEN JENKINS-BATE

*Left above:* Healthy Sunflower Sea Star, San Juan Islands, Washington, 2014 © ED GULLEKSON
*Above:* Sunflower Sea Star killed by Sea Star Wasting Syndrome, Redondo, Washington, 2013 © ED GULLEKSON

African Penguins at Boulders Beach, Cape Town, South Africa, 2019 © ADAM WELZ

Cape fynbos near Cape Point, Cape Town, South Africa, 2021

Fire Daisy; Black Girdled Lizard

King Protea; Southern Double-collared Sunbird on protea

Blue Afrikaner Gladiolus; Cape Dwarf Chameleon on restio

White Spruce advancing into the tundra, western Brooks Range, Alaska, September 2022 © ROMAN DIAL

Burned forest, Kosciuszko National Park, New South Wales, Australia, February 2021
© ADAM WELZ

abnormally high numbers of oxygen-loving bacteria on sea stars' outer surfaces. These bacteria thrive in high-temperature, nutrient-rich waters. In 2013 there was an unprecedented upwelling of such water along North America's west coast, caused by ocean warming, which triggered an algal bloom that increased the nutrient levels in the near-coastal waters where sea stars live. This fed the bacteria, which covered the sea stars in a sort of living film that consumed the oxygen in the water around the sea stars. With little oxygen, sea star cells begin to die and waste away. As they rot, they release more nutrients into the water, driving further bacterial growth—a vicious cycle that the west coast waters show no signs of escaping from.

Giant Kelp forests elsewhere are also in trouble due to warming water. Around the year 2000, Tasmania's kelp forests began to thin out. Pulses of warm water, above the 21 degrees Celsius (70°F) that kelp prefers, were moving through from the north. Photosynthetic cells within the kelp began to warm up, warping and damaging the enzymes within them. Soon they were failing, unable to make carbohydrates to build and support their cell walls, which started to disintegrate. Pale brown patches, like slimy lichen, started to grow across the kelp, and soon the underwater rot reached deep into stems, killing the plants. Southeast Australia has lost almost all its Giant Kelp in the last twenty years. Most of the large fish that used to live in it have moved off, and the sun-exposed seabed is covered in dense thickets of small, shrubby, bright seaweed, now—as off North America's west coast—home to large numbers of sea urchins.

A few Giant Kelp individuals remain in a couple of Tasmania's bays, standing lonely in the current. It turns out that some individuals have the genes to photosynthesize at higher temperatures without succumbing to rot. A small group of scientists has taken samples of these into the lab and has figured out how to grow them. In 2021 divers went down to the seabed with underwater drills to carve out small pockets in the rock in which they placed inch-tall, lab-grown kelp plants. By mid-2022 some of these were already twenty yards long, and very few had died. Human intervention may well be able to bring back part of Tasmania's Giant Kelp wonderland—at least until sea temperatures rise even more.

---

In March 2022, persistent onshore winds blew large volumes of warm water from the central Pacific toward the east coast of Australia. Water temperatures rose sharply along the Great Barrier Reef, between three and four degrees Celsius above the normal maximum. Within days, thousands of acres of the reef, right across its extent, began to pale. The first to bleach were the *Acropora* or staghorn corals, the fastest-growing, best reef builders, which usually grow close inshore and nearer the water surface than other forms. Then more-massive forms began to lose their color. Some corals turned brightly fluorescent as they came under stress—scientists think that some species produce fluorescent pigments to try to reflect excess radiant energy and so reduce their temperatures.

Enormous tracts of the reef began to look sick, but the corals were still alive. Though they'd lost zooxanthellae, they were ingesting nutrients through their mouths. But some types of coral can't survive long on ingested food alone. Within two or three weeks many staghorn corals had starved and died; they're far less able to live without their symbionts than many other types. Shortly after death their complex, branched aragonite skeletons began to break apart and erode.

Corals can recover from bleaching if the water temperature drops back into its normal range before starvation kills their polyps, because the polyps don't usually expel all their symbiotic algal cells. When healthy conditions return, these cells reproduce fast, build up their numbers, and resume their previous function. But research has shown that even if a coral colony survives a bleaching event intact, its growth and breeding rates are stunted for years afterward. A reef can take decades to recover from a severe ocean heat wave.

Although mass bleaching can be triggered by exposure to water just one degree Celsius above the normal maximum that a reef's corals have evolved to tolerate, it appears to be a recent phenomenon, at least in the context of the twenty or so thousand years since the last glacial maximum. The first-ever recorded mass bleaching of the Great Barrier Reef occurred in 1998. The next was in 2002, followed by years of respite before three more in 2016, 2017, and 2020. Scientists noted that each of these years was a strong El Niño year, marked by raised sea surface temperatures in the tropical Pacific, which shaped ocean circulation so that warm water accumulated on the east coast of Australia. La Niña years, defined by lower-than-normal surface temperatures in the tropical Pacific, have always brought cool water to the Great Barrier Reef, giving it a chance to recover from heat.

But the pattern is changing. The year 2022 was a La Niña year, yet warm water still came and bleached much of the reef. There is now so much thermal energy in the surface waters of the Pacific that an onshore wind can override the strong La Niña currents. Climate modelers have predicted that this might happen by 2100, when average global warming reaches three to four degrees Celsius. No serious climate researcher thought that the La Niña barrier would be broken so soon.

Corals are dying rapidly elsewhere too. The vast majority of reefs off the southern coast of Florida have no living stony corals anymore—they've been replaced by soft corals, sea fans, crinoids, and other creatures. Many Caribbean reefs are going the same way.

Corals occupy only one-fifth of one percent of the ocean floor, but they support at least one quarter of Earth's marine species. Recent studies predict that if the planet's average surface temperature increases to 1.5 degrees Celsius above preindustrial levels, 99 percent of coral reefs will be subject to bleaching events so frequent that they will be unable to fully recover. At 2 degrees above the preindustrial average, 100 percent of reefs will enter a death spiral. We are presently on track to reach 1.5 degrees above preindustrial averages in the 2030s.

———

The drop in global temperatures at the Last Glacial Maximum, about twenty thousand years ago, had a profound effect on coral reefs. Temperatures in tropical and subtropical regions were almost six degrees Celsius lower than they are today. Many marine waters became too cold to sustain corals, and with so much water contained as ice on land, the sea level dropped—in some places over a hundred meters—exposing reefs to air.

The Red Sea lost all its corals at this time, but as Earth warmed up in the Holocene, it became suitable coral habitat once more. Because of slow-flowing ocean currents and shallow seabeds, the southern reaches of the Red Sea and its near neighbor, the Persian Gulf, are the two hottest tracts of ocean on the planet. As coral polyps started to find their way back into the Red Sea about nine thousand years ago, floating in from the main body of the Indian Ocean through the Gulf of Aden, they encountered ocean temperatures of about 33 degrees Celsius (91°F). This probably killed many young corals and their algal symbionts, but some survived and, over many, many generations, they evolved the ability to thrive at these extremely high temperatures. They built new colonies and new reefs, and with each spawning event they made their way a little farther north toward the Sinai Peninsula. From there they colonized the narrow Gulf of Aqaba, which runs between today's Egypt, Saudi Arabia, and Jordan and terminates at the southern tip of Israel.

In recent decades the Gulf of Aqaba has been warming faster than the global average. It is now regularly washed through by pulses of water four or five degrees Celsius warmer than its normal maximum, well above levels that cause extreme bleaching and coral death on reefs in other oceans. Ecologists and dive-tour operators were alarmed by these hot pulses when they first appeared—Aqaba's reefs have become world-famous diving destinations, sustaining a large industry—but their effect was, and still is, indiscernible. The corals of Aqaba simply haven't bleached in response to them.

Scientists have taken corals and their symbionts from the Gulf of Aqaba into the lab and experimentally exposed them to different temperatures, carefully observing their responses. Although they appear physically identical to members of their species from the greater Indian Ocean—which stress and die when exposed to warmer-than-normal water—Aqaba corals and their companion algae simply metabolize more efficiently and grow faster when exposed to warmer water. They are healthier because their ancestors' slow journeys through the unusually warm waters around the southern entrance to the Red Sea eliminated genes that made them vulnerable to high temperatures. The mouth of the Red Sea was an evolutionary filter that endowed them with genes that optimized their metabolisms for warm water. In fact, by colonizing the normally cooler Gulf of Aqaba, these creatures have accepted what is to them a suboptimal (though good enough) environment.

As the world's oceans warm up and eliminate almost all stony corals from their habitats, the last refuge on Earth for these remarkable creatures (which first evolved over 160 million years ago) is likely to be in this small, narrow stretch of sea in the Middle East. Stony corals will make their last stand here, represented by these particular populations that were given thousands of years to adapt to high temperatures—time that other corals are fast running out of.

# 8

# Stable / Unstable

The Cape Peninsula is a thirty-mile-long claw of mountainous land that reaches out from the southwestern corner of Africa into the Atlantic Ocean. Dominated by the iconic flat-topped Table Mountain and ringed with bright sandy beaches and spectacular sea cliffs, it's one of Africa's leading tourist magnets. In a regular year over a million visitors go to the peninsula's southern tip, which is preserved within a national park, many borne there in humming, hermetic tour buses that speed through the pristine landscape to a cramped waterside parking lot. Here they join a multilingual procession, excited moblets of people lining up for obligatory snapshots at signs reading THE MOST SOUTH-WESTERN POINT OF THE AFRICAN CONTINENT.

Most tourists file directly back to their vehicles after grinning for the camera. Some stand around for a little while, penguinlike, looking at the cold, heaving ocean, crowded with slick ribbons of giant kelp that jostle in the swell. Passing seabirds dot the air, and a few large, dark-brown Cape Fur Seals lie around on distant rocks. Almost none of the visitors pay attention to the substance of the peninsula itself, the land, perhaps because it serves up none of the clichés of Africa that they've paid to see on their trips. There are no flat-topped thorn trees scattered through golden savanna grasslands here, no sweating jungles, no lurching expanses of desert sand. This rugged landscape could perhaps be said to evoke something of the Scottish Highlands as shown in whisky ads, or some wild bit of the central California coast without its towering groves of redwoods. It's rocky and covered with dense, low shrubby heathland that can at first appear bland but then reveals itself in multiple shades of green and grayish brown, sometimes speckled with flowers.

Charles Darwin, when he landed on the Cape Peninsula in nearby Simons Bay in 1836 during his famous round-the-world voyage on the HMS *Beagle*, wrote that there was "nothing worth seeing here." He immediately hired a horse and carriage to take him twenty-seven miles north to the growing colonial settlement of Cape Town, remarking on the "unconcealed bleakness" of the landscape along the way. A few days later he toured a short distance inland to the winemaking town of Stellenbosch, afterward writing, "There was not even a tree to break the monotonous uniformity of the sandstone hills: I never saw a much less interesting country."

But if Darwin had taken a slow walk across the peninsula and applied his skills as a naturalist, he would have seen through the superficial dreariness of the vegetation and found new species with almost every step. The southwestern corner of Africa is in fact a botanical wonderland, unique in the world.

He would have found fine-leaved shrubs in the same genus, *Erica*, as the famous European heaths. Although most European landscapes have just one or two *Erica* species, here he could have found one or two dozen in a single day, with a dazzling variety of flower forms. Many Cape heaths erupt in tiny, bell-like white, mauve, pink, or purple flowers and transfer their pollen on the wind. Others attract insect pollinators with larger white, yellow, purple, or coral blooms. Still others use bunches of nectar-rich tubular flowers in bright pink, purple, cerise, vermilion, scarlet, ruby, or deep orange to draw in multiple species of small, iridescent sunbirds—the Old World analogues of American hummingbirds—to spread their pollen around.

Darwin would have found a huge range of restios, which at first glance appear to be tough, simple reedlike grasses (or grass-like reeds); they are in fact their own family of plants, which originated just before the extinction of the dinosaurs sixty-six million years ago, when Africa was still joined with the other southern continents to form the supercontinent of Gondwana. Many Cape restios are fine and wiry, barely more than ankle height. Others loom overhead like fluffy dark-green bamboo. (There are a diversity of true grasses here too, but they don't form almost dense fields as in African savannas; they're interspersed among other plants.)

Darwin would no doubt have been intrigued by the proteas, shrubs with long leathery or hairy leaves, many bearing cones and spectacular flowers. (The father of modern scientific taxonomy, Carl Linnaeus, named proteas after the shape-shifting Greek god Proteus because they come in so many different forms.) In spring, proteas festooned with bright yellow, pincushion-like flowers decorate the hillsides. The King Protea's flower is the size of a dinner plate and is ringed by a crown of sharp pink petals. The Silver Protea is a rare treelike species that literally shines; its leaves are sheathed in reflective silky hairs, and in the last yellow light of a Cape sunset they appear to be made of white gold.

Among the ericas, restios, and proteas Darwin could have discovered hundreds of species of geophytes—plants with underground nutrient storage organs like bulbs and rhizomes—including irises, lilies, and orchids, many with outrageously complex and colorful flowers. In the evening an ever-changing spectrum of mysterious, low-hanging perfumes would have filled his lungs. (It took centuries for botanists to discover that many cape plants release airborne sex hormones at nightfall to lure nocturnal pollinators like moths, and many of these gases happen to smell good to humans.)

With some study, Darwin would have found that most of the Cape Peninsula is covered with a unique, hyperdiverse plant community, a vegetation type called fynbos today. (*Fynbos* means "fine bush" in Afrikaans; the stems of its species are too slender or fine to use as timber.) In formal botanical terms, fynbos is treeless heathland defined by the presence of restios and noted for its diversity of proteas, ericas, and geophytes. It is the signature vegetation type of the Cape Floral Region, which occupies the southwestern corner of Africa, extending about 500 miles east of the Cape Peninsula and about 150 miles north. This area is dominated by a Mediterranean climate, cool and rainy in winter and hot and dry in summer, unlike the eastern part of southern Africa, where winters are dry and summer brings rain.

Four other major vegetation biomes (distinctive ecological communities) are found in the Cape Floral Region: forest, subtropical thicket, renosterveld ("rhinoceros veld," a low, bulb-rich vegetation type that historically supported herds of black rhinoceros), and succulent karoo (a hyperdiverse semidesert biome). These often form "islands" within large stretches of fynbos.

Fynbos and these associated vegetation types are so distinctive that botanists classify them collectively as the Cape Floral

Kingdom, one of only six floral kingdoms on Earth. Its plants cover a mere 0.04 percent of our planet's surface, the smallest area occupied by any floral kingdom. (By contrast, the Boreal Floral Kingdom in the Northern Hemisphere occupies about 40 percent of all land.) The Cape Floral Kingdom contains almost ten thousand plant species, over half the number found in the United States but in less than one-thousandth of the area of the United States. Almost 70 percent are endemic, naturally occurring nowhere but within the Cape Floral Region. It has almost seven hundred species of *Erica* versus only twenty-one in the whole of Europe, over three hundred restios, and over three hundred proteas, too. It has the richest geophyte flora of any plant kingdom.

No other temperate region packs so many plant species into so small an area; only the most diverse parts of the wet tropics exceed it. The tiny Cape Peninsula alone harbors over 2,200 species of indigenous plants, far in excess of the 1,400 or so found in all of the British Isles, Darwin's home.

Although the Cape Floral Region's megadiverse status is overwhelmingly due to plants, it supports fascinating animals too. It has a variety of strange frogs, such as rain frogs in the genus *Breviceps* ("short head"), which have short legs, almost spherical bodies, and faces like morose gnomes. They can't hop—they only walk—and their entire life cycle takes place on land; they dig small tunnels and lay their eggs in a jelly mass underground, inside which their tadpoles develop into frogs. (*Breviceps* will drown if dropped in water.) The Cape Mountain Toadlet is found only in a small area just north of Simon's Bay and—almost uniquely among frogs—has no vocal equipment or eardrums. Males and females find each other by habit in a couple of small bogs. On a few nights every year, hundreds pile up together in slime-slicked mass orgies to release their eggs and sperm.

The Cape Floral Region hosts the highest concentration of land tortoise species anywhere on Earth, and a high diversity of other unusual reptiles. Black Girdled Lizards, which live only on the Cape Peninsula, look like six-inch-long alligators that have been dipped in blackboard paint. It's thought that black skin allows them to efficiently absorb solar energy even on the cool, ocean-fogged days that the peninsula often experiences (closely related species that live a short distance inland are paler brown). The floral region's rivers contain diverse species of rare fish. Although the Cape Floral Region has very few unique, endemic bird species (six or seven, depending on who's counting) a couple of these play vital ecological roles by pollinating fynbos flowers: the Cape Sugarbird, a long-tailed brown bill with a thin, downcurved bill, appears to be a very old species. Its song is a premelodic jumble of staccato notes interspersed with metallic groans and pops, and it feeds almost exclusively on the nectar of protea flowers. Another nectar-feeding fynbos endemic, the small but spectacular Orange-breasted Sunbird, pollinates a wide variety of flowering species. It's not related to the hummingbirds of the Americas, but it's evolved bright, iridescent plumage like them and fulfills a similar ecological role.

Cape plants found their way into European horticulture in the centuries before and after Darwin's visit, and from there they were spread around the world. The forebears of garden varieties of lobelia came from here, as did many colorful gladioli, ixias, sparaxis, gazanias, lilies of the Nile (*Agapanthus*), naked ladies (*Amaryllis*), and fragrant freesias. Many have been hybridized and selectively bred into forms that no longer resemble their wild ancestors, like the ubiquitous window-box geraniums, which are derived from Cape *Pelargonium* species. White Calla Lilies, called Arum Lilies in South Africa, come from soggy areas of the Cape fynbos.

Some Cape plants are now noxious weeds elsewhere, like Ice Plant, an invasive creeping succulent that suffocates Californian coastal dunes. It hails from Cape Town, where it's known as the Sour Fig. Bitou Bush, a shrubby yellow daisy, has taken over lengths of the Australian coast.

So the fynbos is the flagship vegetation type of a strange and extraordinarily diverse floral region, a botanists' Shangri-La, a horticulturists' treasure trove, and a naturalists' Pandora's Box. But botanists have struggled to understand why it contains so many species, because at first glance this habitat can seem like an awful place to be a plant.

Many Cape soils are extremely nutrient-poor, lacking in nutrients such as nitrogen and phosphorus. They're barely more fertile than beach sand, and they're often highly acidic and hydrophobic, too; water runs off them before it runs into them, so it takes a lot of heavy rain before water penetrates down to plants' roots. Also, the Cape summers are long, hot, and windy, with little rain; the soils dry out. Very little of the organic matter (like dead leaves) that falls from plants gets broken down into nurturing compost because so few insects can survive on the little moisture available. Some fynbos plants produce about the same amount of living material each year as desert plants. All this means there's relatively little nutrition available for herbivores, which in turn means fewer carnivores than in many other African ecosystems.

And then there's fire: every decade or two, usually in the dry summer, the average acre of fynbos burns down. Lightning, sparks from a rockfall, or, nowadays, a wayward cigarette or arsonist's match can give birth to cartwheeling flame-wraiths leaping through the scrub that build into roaring, pulsing walls of fire thirty or forty feet high and miles across. They can burn for days or weeks, sending up waves, towers, mushroom clouds of pale gray-brown smoke that top out well above ten thousand feet and obscure the sun, reducing it to a sharply defined tobacco-orange polka dot in the sky. Big fires can cover tens of thousands of acres and send the sour smell of burned plants and animals a hundred miles downwind.

They leave so little color in the landscape that the blue sky looks garish, artificial, wrong. After a big fire, almost everything aboveground has been vaporized, exposing vast areas of thin, pure pale-gray sand overlaid with wind-swirled eddies of off-white ash so light that gravity barely holds it. The inner skeletons of some large shrubs persist, looking like charcoal antlers bent away from the prevailing winds that shaped them while they grew. A few corpses of animals lie on the sand, perhaps a curled-up, asphyxiated Four-striped Grass Mouse or the lifeless shell of an Angulate Tortoise surrounded by a neat ring of black and red-brown scutes; the fire's heat caused them to fall off, revealing the animal's domed skeleton beneath.

But despite its species having to deal with extremely low-nutrient soils, desertlike summers, and regular conflagrations, the fynbos has become extraordinarily diverse. How? After decades of research, scientists think they have some answers: its richness results at least in part from frequent fire, the Cape's broken landscape, and a climate that's remained stable for a very long time.

---

Darwin was blind to the riches of the Cape fynbos vegetation, but this doesn't invalidate his fundamental insights about evolution. As outlined in chapter 1, he noticed that species evolve because their individual members are slightly different from each other. Those that are more fit are more likely to survive and reproduce than less fit individuals. Over time, the characteristics of the fitter individuals will become prevalent in the species, and it will evolve.

The main source of individual differences is genetic mutation, i.e., changes in DNA sequences that occur during cell division as a result of errors in DNA copying; specifically from errors that occur during the creation of sex cells (eggs or sperm). When a male and a female sex cell combine to make a new organism, their mutations can be expressed in the organism as new characteristics, different from those of its parents.

Each generation thus offers an opportunity for mutations to occur. The more generations a species passes through, the more chances it gets for genetic change, so the shorter a species' generation time, the faster it can evolve.

Fossils indicate that fire and fire-adapted plant species were present in the Cape at least eighty-one million years ago, in the Cretaceous period. Fynbos plants have had a very long time to evolve variations on their two main strategies for surviving fire, which are sprouting and seeding. Sprouter species regrow after fire from protected buds. They put resources into growing a large underground structure stocked with nutrients, like a bulb or rhizome, or build thick, fire-resistant bark that protects dormant buds buried deep within their thickest stems. After a fire has passed, they sprout new stems and leaves from their underground storage organs or from their protected buds.

Adult seeder plants, however, are killed by fire. Their species persist in seed form, and they've evolved numerous ways for their seeds to survive blazes and germinate once the flames have died down. Many proteas are serotinous, keeping their seeds in tough cones, which open after exposure to high fire temperatures and spread the seeds on the now-cleared sand, where they can benefit from the nutrients in fire ash. Other seeders shroud their seeds in tasty nutrients to entice ants and rodents to take them into colonies and burrows underground, where the seeds are safe from combustion and are triggered to germinate by the warmth of a fire passing above them or chemicals in smoke.

Because adult seeders are killed by fire, each fire forces a new generational cycle to begin. Thus the fynbos's seeder species can evolve faster than they otherwise might have, making new species more rapidly than if all of the fynbos was composed of sprouters.

Fire also creates an enormous number of ecological niches that open up and narrow down as fynbos habitats regrow after fire. An acre of fynbos thus in effect hosts many different habitats that emerge and recede in a roughly predictable sequence through the fire cycle. The sequence goes like this: just days after a typical midsummer fynbos fire, the first so-called fire ephemerals emerge from the scorched sand. These are plants like the bright red Fire Lily and yellow Fire Daisy, which flower only immediately after fire and aren't visible for long. (Many fire ephemerals are small plants that struggle to compete with larger plants for nutrients, light, and pollinators, and their rapid postfire emergence triggered by chemicals in smoke, allows them unimpeded access to these resources.)

Shortly afterward restios and some bulbs begin to slowly resprout, and months later, as the winter rains begin, many seeder species, including ericas and proteas, germinate. The race to acquire nutrients and water ramps up with the next onset of spring, when annual daisies and many bulb species erupt into kaleidoscopic swaths of color. By midsummer a year after the fire, the dry season has set in, and seedlings that didn't succeed in getting their roots deep enough die off. A wide variety of low native grasses are establishing themselves throughout the landscape. At this stage the vegetation is open, no more than knee height, and home to open-country, ground-dwelling birds like larks and lapwings, as well as slowly recovering rodent and reptile populations.

Over the next few years restios and larger sprouter species progressively dominate the smaller early-emerging plants. As each winter rainy season drives a burst of growth, the vegetation becomes denser and taller. After five years or so, a broken shrub layer made up mostly of larger proteas brings a new dimension to the fynbos, creating new habitat for songbirds as well as shaded patches where insects and shade-tolerant plants can thrive. The shrubs will slowly become ever more substantial and the underlayers ever more dense, with abundant food and shelter for small mammals and the snakes and mongooses that prey on them. (Although many fynbos plants have evolved techniques to pull nutrients from extremely poor Cape soils, like ultra-thin roots and symbiotic relationships with soil microorganisms, their growth is ultimately nutrient-limited. Fynbos shrubs can't grow into large trees.)

Fire and the ecological succession that follows in its searing wake means that each square yard of fynbos terrain can host a remarkable number of species. Species take turns to occupy significant space, depending on which stage of the cycle they have evolved to emerge in. Many species spend much of the cycle as dormant seeds or tiny bulbs, taking up hardly any room. In the same way that a military submarine can carry three times as many personnel as it has beds available because two-thirds of its crew are out of bed and on shift at all times, the fynbos can host more species than you might expect because its plants are top-notch "hot bunkers."

The second major factor that contributes to the diversity of fynbos is the Cape landscape, which has a long and complex geological history. The region is defined by many old coastal mountain ranges, collectively known as the Cape Fold Mountains. Their peaks are separated by deep valleys, and the region has many different rock types, which break down into many different soil types. This variety of altitudes, aspects, and soils gives rise to an extraordinary number of microclimates and potential niches for plants to diversify into. Yet the overall climate across the Cape Floral Region is generally similar regardless of location; hot, dry summers and cool, rainy winters. If a few seeds of a mountaintop plant get carried to another peak fifty miles away in the guts of a bird, chances are that the seeds can germinate into plants that will survive, reproduce, and become the sort of isolated population that turns into a new species. If another, similar plant lands on the peak some generations later and begins to compete with the earlier arrival, the varied topography will ensure enough niche space for them to differentiate into; competition will lead to diversification, not the obliteration of one species by the other.

Mountains create abundant opportunities for organisms to be reproductively isolated in similar climates; the Cape peaks are like a chain of islands that can create lots of new species relatively quickly.

The third major factor, the stable climate, is explained by the Cape's temperate location—it's been at moderate latitudes for a long time, and was not sterilized by ice sheets during the Pleistocene—and the cool Benguela Current, which flows from the sub-Antarctic toward the Cape and then northward up the west coast of Africa. The Benguela is the eastern arm of the colossal South Atlantic Gyre, which rotates counterclockwise around the South Atlantic. When South America separated from Antarctica around the beginning of the Miocene epoch, some twenty-three million years ago, this set the Antarctic Circumpolar Current strongly into motion, which in turn pushes the South Atlantic Gyre around. The consistent Benguela is a strong influence on the climate of the southwestern corner of Africa, taking the edge off hot summers and helping to generate regular, reliable winter rains in the Cape; it's a vital driver of the Cape's very old Mediterranean-type climate.

Although Cape has likely experienced reliably wet winters for more than twenty million years, its climate hasn't remained *completely* stable in that time. It's varied a bit as the global climate has varied between glacial and interglacial periods, for example. The winter rains have increased or decreased in intensity and the size of the area of South Africa that they've fallen over has waxed and waned.

These climate changes have encouraged speciation by allowing or forcing organisms to move into new areas. But the extent and speed of the changes has been moderated by the Benguela Current, which means that few species have experienced dangerously fast climate change. It seems that the Cape's climate has changed at a Goldilocks pace for millions of years: fast enough to drive the evolution of new species, but not so fast as to drive them extinct.

Even though the Cape Floral Region's soils are weak, its summers are punishing, its annual rainfall is often low, and fires regularly sweep across it, its vegetation nonetheless is hyperdiverse because its extinction rate is hyperlow. Climate stability counts.

———

Global heating is often expressed in terms of an increase in temperature since the beginning of the Industrial Age, usually considered to be the period from 1850 to 1900. As I write this in early 2023, the scientific consensus is that Earth has warmed about 1.1 degree Celsius since then, and that it'll reach 1.5 degrees above the preindustrial temperature by about 2040.

This statement is often misinterpreted, though. It does not mean that temperatures everywhere on Earth (or throughout the biosphere) have gone up by 1.1 degree since humans started burning fossil fuels on an industrial scale, and it does not mean that we can expect our local daytime maximum temperatures to go up by 1.5 degrees, for example from 25 to 26.5 degrees Celsius (77°F to 80°F).

What it means is that the *average* of surface temperature readings taken across the globe has gone up by 1.1 degree Celsius. (Surface temperature in this context is not the temperature of the actual land or ocean surface; it is the temperature of the air about one meter above the land or ocean surface, usually taken in a meteorological instrument shelter.) Temperature change has not been even across Earth—far from it! Average surface temperatures in some places have gone up significantly more than 1.1 degree, and in other places they've even declined. The same with minimum and maximum temperatures—these have increased much more than 1.1 degree in some places, and even declined in others.

Although $CO_2$ is a well-mixed gas, meaning that it diffuses rapidly in the atmosphere and atmospheric $CO_2$ levels are very similar no matter where on Earth you measure them, thermal energy is not evenly distributed throughout the atmosphere, land, and oceans; it's notably uneven, and the oceans have a lot to do with this.

Water can hold far more thermal energy per unit than air. It also absorbs thermal energy more effectively and holds on to it more strongly, which is why more than 90 percent of the thermal energy added to the biosphere by human-caused global heating is held in the oceans. The thermal properties of water also explain the significant influence that oceans have on the temperatures and climates of the land, especially coastal land and islands.

The Galapagos archipelago, located about six hundred miles off the coast of continental Ecuador in the eastern Pacific Ocean, is famous for being key to Darwin's development of the theory of evolution. This small group of volcanic islands has about eighteen species of closely related songbirds often called Darwin's finches (scientists don't agree on exactly how many species there are, and they are not actually true finches, but in the tanager family). Darwin and a couple of his fellow travelers shot numerous specimens of these birds when they visited the archipelago in 1835, and in 1837, after arriving back in England, Darwin gave them to the ornithologist John Gould.

Gould soon realized that the "finches" from the Galapagos were not all of the same species. Some had very large, thick bills, others thin, and yet others in between. The specimens could be grouped into distinct species, and those specimens of the same species all came from the same island, so each island, therefore, had its own unique types. He noted that these birds were only known from the Galapagos but were nonetheless similar to species from South America, the nearest continent.

Darwin then surmised that all the Galapagos birds had evolved from a single ancestor, which had somehow found its way from mainland South America to the islands, and then evolved into separate species, each species well suited to a particular lifestyle. "Seeing this gradation and diversity of structure in one small, intimately related group of birds," he wrote, "one might really fancy that from an original paucity of birds in this archipelago, one species had been taken and modified for different ends."

Darwin's finches have become a leading example of how numerous species can arise from a single ancestor, and the Galapagos have become a hotbed of research into evolution. Scientists come here to work not only on the finches but on an array of other unique species like giant tortoises (each species also native to a single island) and strange animals like marine iguanas, which eat seaweed underwater. The islands are a UNESCO World Heritage Site, and their wildlife draws over two hundred thousand visitors annually, as well as a never-ending stream of documentary film crews.

The Galapagos's special species are, however, extremely vulnerable to climate change. The islands straddle the equator and receive a lot of direct sun. They're small, often with little habitat diversity—species don't have many places to move to if their local habitats and climates change. Many species, like the finches, have evolved into narrow ecological niches. As specialists rather than generalists, they should theoretically be among the first to run into trouble as temperatures rise; in fact, early climate studies predicted that many Galapagos species would be at serious risk of decline by 2020.

But although some species are being threatened by direct human pressure—overtourism—and invasive species brought by people, very few are yet showing any sign of endangerment due to climate change. Even though the atmosphere covering the Galapagos can now hold more thermal radiation—like the atmosphere everywhere—it turns out that the ocean around the islands has in fact got half a degree Celsius colder in the last decade. The Galapagos archipelago, like the southwestern Cape, is lucky in its location.

Since the 1980s scientists have noted a large, persistent patch of cold, nutrient-rich upwelling water on the western side of the Galapagos. It feeds a huge abundance of fish and supports the islands' seals and seabirds. But no one knew what generated this cold patch until very recently, and what made it so robust. In 2022, using decades of observations from satellites and floating instrument stations, scientists produced the first map of an influential current that explains/creates the cold patch.

As Earth spins, it creates an ocean current called the Equatorial Undercurrent, a fast-moving stream that shunts water west-to-east about a hundred meters below the surface of the equatorial Pacific Ocean. The Equatorial Undercurrent is far colder than the surface layers, and for most of its length it barely mixes with the surrounding water. But the Galapagos Islands are volcanic; their landmasses are the tips of giant extinct volcanoes, massive cones of igneous rock rising from the sea floor. As the Equatorial Undercurrent rushes toward the islands, it runs into the sides of these volcanoes, which push its cold water to the surface. When the water contacts warm air, it draws thermal energy from the air around the islands, lowering and stabilizing the islands' surface temperature.

While collecting the data that allowed them to explain and map the Equatorial Undercurrent, oceanographers found that the current is moving south at a rate of about a kilometer per year. Ten years ago the center of the current, its coldest part, passed just to the northern side of the middle of the archipelago. Now it runs right through the middle of the islands, bringing more, colder water to the heart of the Galapagos, which explains the temperature drop.

Most ocean currents are driven by temperature differences between different parts of the ocean. But the Equatorial Undercurrent's motion stems almost completely from the rotation of the Earth, which means that it is highly unlikely to slow down or stall as global temperatures rise. But, since no current is an island, other nearby currents may nudge the Equatorial Undercurrent on a different course as they move in response to global heating.

The Galapagos are effectively shielded from rising air temperatures for now. But because the islands occupy such a small area, it would take just a small shift of the Equatorial Undercurrent north or south for them to escape its protective stream. Without it they will be at the mercy of the heating air, and many of the unique creatures that inspired perhaps the greatest scientific breakthrough of the last millennium will disappear.

In the late summer of 1991, I visited Utqiagvik—then named Barrow—the northernmost human settlement in Alaska, on the edge of the Arctic Ocean. Almost everything about the place looked wrong and felt wrong to me. I stood on the beach that ran along the edge of the tiny town. The pebbly sand was dark, almost black, unlike any beach I'd seen before. The sky was a bright, even pink-orange; all of it, not just the bit near the low sun, which wasn't setting but moving in circles around the horizon, making it light twenty-four hours a day. The sea wasn't blue but white, crowded with sea ice. And the land, covered in tundra vegetation, was flatter and more uniform than I'd thought possible, with a more distant horizon than I'd thought possible. (The nearest trees were about 250 miles south, within the massive Brooks Range, through which ran the tree line, north of which trees could not grow because conditions were too extreme.) But even though I could see for miles and miles and miles I hesitated to walk more than a couple of hundred yards from the town; there were no natural landmarks anywhere. I felt like I could get totally disoriented, utterly lost, in this place where you could see forever.

The area around Utqiagvik wasn't strange to most of its human inhabitants, Iñupiat people whose ancestors had been here for over fifteen hundred years, during which time they'd figured out how to live in an ecosystem that does everything in extremes. The summer buzzes with life; the sun doesn't set for more than two months, and every living thing seems to be growing and breeding as fast as possible. The tundra plants grow and insects fill the air, sustenance for millions of migrant birds—sandpipers, ducks, geese, and songbirds—that breed everywhere. Lemmings scamper across the land, providing food for Arctic Foxes and Snowy Owls, while fish fill the estuaries. But winter is dead, frozen, and extremely dark. The sun doesn't break the horizon for sixty-five days. Almost all the birds leave, the insects die, and the plants stop growing. Most of the land mammals hibernate.

Even though high Arctic summers and winters are so extraordinarily different, the passage of the seasons here has been highly predictable for as long as people have lived at Utqiagvik. Its residents have learned to read their environment and hunt different animals, from Bowhead Whales to seals to fish, at very specific times of year. But now their ancient knowledge is becoming increasingly irrelevant and it's becoming possible to imagine vast areas of tundra in northern Alaska becoming shrubland or even forest; although trees can't walk, they are marching toward Utqiagvik.

In recent years scientists have found White Spruce saplings in northwest Alaska advancing into zones that they thought could never support trees. Frozen soils in many Arctic regions are melting and letting in tree roots for the first time in tens, maybe hundreds, of thousands of years. Stronger winds are blowing White Spruce seeds further than before, over the mountains. A 2022 study used aerial photos and satellite images to show that White Spruce trees are moving north through the Brooks Range at an average of four kilometers per decade, faster than they moved to recolonize newly ice-free parts of North America after the Last Glacial Maximum.

Not only are White Spruce youngsters speeding toward the North Pole, but they're growing faster, taller, and healthier than their parents not just because of general warming but because they're changing their own immediate environments. Because they stick up from the land, blown snow accumulates around them, and when it melts, the trees get more water than they otherwise would. Trees are darker than tundra vegetation, so they convert more solar radiation into thermal energy than tundra does, warming the area around them. The warmer the soil around their roots becomes, the more nutrients are liberated by microorganisms, which then metabolize faster and further warm the soil, liberate more nutrients, and grow more microorganisms; a self-amplifying (and, for trees, virtuous) cycle.

Fast-changing tree lines are only one of the undeniable signs of global heating visible in the Arctic, where average temperatures have gone up by about four degrees Celsius since the start of the Industrial Revolution—about four times the global average. Habitats are changing fast all around the region. Areas of tundra are being invaded by shrubs and trees. Forests are drying out, burning, and being replaced by grassland. Permafrost is melting, changing the soil and everything that lives in and on it. Unlike the Cape of South Africa, where the thermal inertia of ocean water has helped to stabilize the climate, here the ocean is a vital part of a self-amplifying cycle that's pushing temperatures up.

There is no land around the North Pole, unlike the South Pole, which is of course where the Antarctic continent is currently located. But there is a northern ice cap, a vast sheet of ice that floats on top of the Arctic Ocean. It grows in the cold and disturbingly dark Arctic winter and shrinks in the equally disturbingly bright summer.

Floating ice topped with a layer of snow is extremely good at reflecting radiant energy from the sun back into space. By contrast, bare, iceless ocean is extremely good at absorbing solar radiant energy and transforming it into thermal energy. So when the Arctic ice cap naturally melts back from its edges during the northern summer, the Arctic Ocean acquires increasing amounts of solar energy. Rising global air temperatures mean that slightly more Arctic sea ice is melting every summer than before. As more ice melts because of the warming air above it, the ocean loses more of its reflective shield and gains more thermal energy, which melts the ice cap from below; another self-amplifying cycle that scientists call Arctic amplification.

Arctic amplification would not occur if the North Pole was covered by land as the South Pole is. The phenomenon is largely a result of water's significant thermal inertia. Recent decades have seen rapid year-on-year shrinking of the area covered by summer sea ice; a study published in June 2023 in the prestigious journal *Nature Communications* predicts that the Arctic Ocean may be completely ice-free as soon as the 2030s, and that even if we immediately reduce fossil fuel emissions, we will not be able to prevent the total loss of summer Arctic sea ice. It will be the first major climate change tipping point that we'll pass, with global impacts that cannot yet be fully understood.

The climate of the Galapagos is barely changing. The Arctic is transforming before our eyes. But what of the southwestern corner of Africa and its Cape Floral Kingdom, an excellent example of how long-term climate stability can support the flowering of life?

Since Darwin landed on the Cape Peninsula, Cape Town has grown into one of Africa's richest cities. Its modern history begins earlier, though. In 1652 the first permanent European presence was established here by Dutch colonists, who built a small settlement on the shores of Table Bay, a stopover on the prosperous shipping routes between Europe and Asia. In time, the British took over. Early European settlers developed some agriculture here—the region has been famous for good wine since shortly after 1688, when fleeing French Huguenots arrived—but because the soils of the Cape are generally poor, much of the region remained unplowed. Cape Town continued growing, and by the early 1900s the city had become a large urban center, with hundreds of thousands of residents, yet agricultural expansion proceeded relatively slowly.

This changed in 1930, when the South African government restricted wheat imports and triggered a boom in wheat farming (which was helped along by new synthetic fertilizers). Within a few years most of the renosterveld was gone, along with huge areas of lowland fynbos. Herds of cattle and sheep expanded, degrading more natural habitat. In 1948 the notorious apartheid government came to power in South Africa, and heavily restricted Black people from living in Cape Town; only so-called Coloured (mixed race) and white people were encouraged to live here. By 1980 Cape Town had about 1.6 million residents, booming industrial areas, and ever-expanding suburbs. Many of city's lowland habitats were being covered in roads and buildings, and numerous native plant species were highly threatened or extinct. In the rural areas of the region, more roads and increasingly mechanized agriculture broke up and consumed yet more natural landscapes.

In the late 1980s and early 1990s two consequential things happened at once: a massive, multiyear drought enveloped much of South Africa, and the apartheid system began to fall apart. Huge numbers of rural people who had formerly been forbidden to live in Cape Town streamed into the city. (By the mid-1990s, three thousand people per day were moving in; the city's population in 1994 was estimated at 2.3 million.) Massive shantytowns spread over most of the city's remaining lowland natural habitats. No one knows exactly how many people live here today; mainstream estimates are about five million.

Some areas of the Cape Floral Region are conserved inside nature reserves and national parks. But hundreds of fynbos plant species are now endangered, many of them down to a few individuals in small patches on the sides of freeways, in pocket parks, or on trash-strewn land on the city's edge.

Some of the world's best botanists and climate scientists are based in Cape Town, and they've been hard at work trying to understand the region's future. Long-term climate models predict that the Cape, along with the western half of South Africa, will see progressively hotter temperatures and less rainfall as the century unfolds. So far, weather records have generally confirmed this trend.

There is already evidence that higher temperatures are killing off populations of fynbos plants. In 1966, a botanist laid out fifty-six study plots in the fynbos near Cape Point, not far from where Darwin landed 130 years before, and logged every plant species within them. Other botanists resurveyed the plots in 1996 and 2010, and found that many species had disappeared. Losses were particularly evident in plots that had burned just before dryer-than-normal summers. Very young plants (resprouting or germinating after fire) are particularly vulnerable to drought, and the scientists found that between 1966 and 2010, periods of consecutive hot and dry days had become longer, and the average maximum daytime temperature had risen by 1.2 degree Celsius. Some plots had gained a few new species, and the researchers found that these were species with greater tolerance for high temperatures than those that had gone. Warming is already harming fynbos species, even within the best-preserved portions of national parks.

Models indicate that as climate breakdown proceeds, hundreds or thousands of species of lowland fynbos plants will not be able to survive in their current habitats. They will have to move upslope, to cooler areas, but this is far easier said than done because many fynbos species lack long-distance seed dispersal methods; their seeds don't travel great distances on the wind, for example. Even under ideal circumstances, few will be able to send their seeds upslope fast enough to escape rising temperatures.

And circumstances are not ideal. Plants and small animals have their potential paths to friendly microclimates blocked by shantytowns, expanding wealthy suburbs, and freeways. Beyond the city, intensive farmlands have inflicted perhaps even more damage on natural systems. Nitrogen fertilizers have enriched the soil, making it toxic to fynbos species that have spent millions of years fine-tuning their physiologies to extremely low-nutrient substrates. And wave upon wave of invasive trees and shrubs from all continents bar Antarctica, transported here by humans, are spreading like an infectious disease across the destabilized Cape, suffocating thousands of acres of the fynbos that remains. (You could of course write sentences like these about many other places in the world.)

Although Darwin didn't appreciate the fynbos, many contemporary conservationists do. They're planning conservation corridors so that those organisms with the ability to travel across the land-scape can get to places where they can persist in the future. They're working with almost no money to grow endangered plant species in nurseries and greenhouses in the hope that they can one day—who knows when—be replanted into the wild. Perhaps if the Benguela Current continues to moderate rising tempera-tures and keep the winter rains coming, and if energy-trapping emissions soon cease to pollute the air, these human efforts will mean something. Perhaps, with a lot of luck, the species of this incredible biome will continue to enchant your descendants and mine.

9

# Conclusion

W e're only just beginning to break apart the stable climate system gifted to us by the Holocene epoch but, as we've seen through these pages, the natural world is already sustaining significant harm. Species that have survived for millions of years are suddenly being forced beyond their limits. Ecosystems are being scrambled. Large Earth systems are being pushed toward any number of tipping points, which can trigger cascades of further, runaway transformations that will irrevocably disrupt and diminish the entire biosphere.

I can't overstate the consequences of this. Although there are many opportunities to slow and avert species extinction, soften many of the worst blows of climate breakdown, and create virtuous, self-reinforcing cycles of action that improve the prospects for all life, we must act quickly and decisively to seize them. The longer we fail to make meaningful changes to our economies, our political systems, and our moral frameworks, the fewer good options we will have for curtailing the extraordinary damage being done by climate breakdown.

It's perhaps understandable that the stories we hear and tell ourselves about the climate crisis tend to focus on its dangers for human societies and economies—what big storms, severe droughts, and rising sea levels are doing and will do to us. But *Homo sapiens* is a generalist, adaptable species. Some human societies will be able to cushion themselves against the impacts of climate breakdown, at least in these beginning phases. Many, if not most, wild species— already depleted by human destruction of their habitats, invasive species, and so on—are significantly more vulnerable and more immediately threatened.

I'm acutely aware of the peril that wild species are in because I've been observing and studying non-human life for some forty years, both for professional reasons and to satisfy my nature addiction. I've seen species vanish from places where I knew them to be common and other species newly establish themselves far out of their historical ranges. The scientific findings about climate breakdown make sense to me because they line up with my own experience; I have personal baselines against which I can assess change, especially in South Africa, where I was born and raised.

As a teenager in 1980s Pretoria, which lies at the interface of highveld grassland and bushveld savanna habitats, I caught and kept snakes, lizards, frogs, and fish, and enthusiastically watched birds. I wandered around vacant lots and neglected natural areas on the city's ragged edges with snake-obsessed buddies; we spent days turning over rocks and rusty sheets of corrugated iron in search of reptiles, and evenings paging through books to learn their scientific names. I came to know my local birds intimately, like the rival pairs of Crested and Black-collared Barbets that fought noisy battles for nest holes outside my window, and the tiny Willow, Icterine, and Olive-tree Warblers that flew annually from Europe to the tall Sweet Thorn tree out back, where I watched them hunt bugs as I added them to my bird lists.

I eagerly volunteered my labor to local biologists, sometimes traveling hundreds of miles into the bush with them to seek out rare lizards and snakes, and often helping them trap birds for ornithological research.

My role as an unpaid scientific assistant led to many adventures and new experiences. One evening we went to a large reedbed on the city's northern side, where the European Barn Swallows that migrated to our region every summer roosted at night. They began to gather in the air above the roost about half an hour before sunset, first a few groups of hundreds, then many thousands, and, before you knew it, a dark cloud of literally hundreds of thousands or a million swallows was whirling overhead. Their collective twittering was a screaming roar, made more overwhelming by the sharp smell of ammonia rising from the thick, off-white layer of slimy droppings that coated the ground around us.

Then, around sunset, they began to descend into the reeds, not by gracefully gliding down, but by pulling their wings in and dropping like feathered hailstones into the vegetation; a dense, disorienting storm of falling birds. It lasted maybe ten minutes and then the sky was clear. Our long mist nets, stretched between poles planted at the edge of the reedbed, were packed with hundreds of swallows, which we untangled to weigh, measure, and band with individually numbered aluminum leg rings.¶¶ (We would release them unharmed the next morning.)

Almost all the birds in our nets were unbanded—they'd not been caught before—but a handful were valuable retraps, two of which carried foreign rings, one from Estonia and another from Israel. Catching these birds was like finding a message in a living bottle, and I've never forgotten it. The tiny rings around their fragile legs were proof that they really could fly across the planet, even though the average Barn Swallow weighs less than 20 grams—about two thirds of an ounce. Not only that, but these very individuals with their bright, alive eyes and silky-soft feathers had been held by real people somewhere behind the imposing Iron Curtain and in the conflict-wracked Middle East; real people who cared about them just as we did.

Finding these birds in isolated, pre-internet apartheid South Africa was amazing to me, and it stands out as an example of how wild creatures offered me an escape from the conformist, white supremacist society of my youth. Wild species generated interest, demonstrated beauty, and inspired engagement with new worlds. Mainstream white society had little to offer me but a rotting smorgasbord of racism, militarism, and uptight conservative Christianity, heavily seasoned with gray, deathly boredom.

I got out of Pretoria as soon as I was done with high school. Since then I've traveled, lived, and worked in other parts of South Africa and many other places, including the United States, Taiwan, India, Madagascar, Brazil, northern Europe, and numerous African countries. In all these—even in depressingly nature-poor Great Britain, the small, sodden island state where I eked out a threadbare

---

¶¶ At that time few ornithologists were interested in climate breakdown—the issue was barely on ecologists' radar screens—but the data gathered by bird-banding projects like this one are now vital in understanding how migration routes and timing are changing with the climate.

living for over two years—I've sought out wild nature. I now live in Cape Town, surrounded by mega-diverse fynbos habitats and rich kelp forests.

While many people who studied biology at university with me have gone into high-tech fields, working in DNA labs to understand evolution or analyzing satellite data to chart ecological change, I've become an unfashionable, old-school naturalist: I'm interested in all wild species, everywhere. I'm concerned with knowing their names, identifying them, and observing them firsthand in their natural habitats. I also find great meaning in keeping them alive: my writing office has become an ark for threatened fish species, which I breed in numerous tanks arrayed around my desk, and I grow native plants from seed I've collected nearby.

I feel that people who don't share my nature fascination, who confine their attention solely to our species, are missing out on most of life. Being aware of the wild creatures around me adds what I can best describe as an ever-expanding extra dimension to my world. I can be working at my desk, hear a songbird, and within seconds know its species and whether it's relaxed or stressed, looking for a mate, begging from a parent, staying in touch with its flock, or warning of a nearby predator—perhaps the local Spotted Eagle-Owls are roosting in the Milkwood branches outside, or the neighbor's roaming housecat is trying to kill the last remaining Cape Skinks in my garden? Other species give me different clues: a passing butterfly catches my eye through the window—it's the first Cape Autumn Widow of the season, so biological summer must finally be over. Local fynbos plants reflect the changing climate through the timing and fecundity of their flowering: Were the winter rains early or late, weak, normal, or large? The more I pay attention to them, the more the plants can tell me.

Some acquaintances think that my knowledge of nature and ability to detect wild animals (especially birds) are extraordinary

("How did you know/see that?!"), but I'm nothing special. My preindustrial ancestors doubtless knew and perceived the natural world far better than me; they would not have survived otherwise. Although many people alive today have never exercised these abilities and are thus nature-blind and ecologically illiterate, humans are built to tune in to nature.

I've traveled in some of the planet's most spectacular wild places, but some of my best encounters with nature have been in large cities. When I lived in New York I watched a Red-tailed Hawk jink along a bustling Fifth Avenue sidewalk, waist height between shoppers, to hit an unwary pigeon and eviscerate it for lunch. I've seen Peregrine Falcons circle high in the beams of the September 11 memorial lights over lower Manhattan to prey on disoriented night-migrating songbirds, sat with a Northern Snapping Turtle as she patiently dug a nest and laid her eggs just inches from a busy sidewalk in the Bronx's Van Cortlandt Park, gazed at thousands of large, humming Green Darner dragonflies migrating together over the noisy streets of downtown Brooklyn, and so, so much more.

Wild species provide entertainment, but their behavior—or just the simple fact of their presence or absence—also gives me meaningful and practically relevant information about the places we inhabit. I'm aware of being surrounded by not exactly friends, but accompaniers—other beings who have their own aims and motivations, but whose fates brush up against me and unfold in my company.

And because there's always more to find out and connect with in the greater-than-human realm, I am never bored.

---

The word that comes to mind is *squander*. We built our human societies within a natural world just past the peak of its full flowering but still thriving, resilient, and diverse. Despite scientists' impressive ability to stare deep into space, they've not yet found a planet with complex life like ours. Its richness and beauty is nothing short of a miracle, though not a miracle created by an all-controlling God but one that emerged via the processes of energy and matter being shaped through evolution over billions of years. Now we're ripping it up, piece by piece, and tossing it in the trash with relentlessly increasing speed.

I think of natural species as victorious answers that have arisen through evolution in response to life's most fundamental challenges: survival and reproduction. Each species is not, however, a single answer—it is an answer made of answers, a fractal-ish assembly of answers. A fish species, for example, is a successful answer to the problem of surviving and reproducing in water, but it depends on components like gills, which answer the question of how to obtain oxygen underwater, and fins, which answer the question of how to move around in the aquatic realm. Different tissues and cell types within these components are answers to particular physiological challenges, and they're made of specific molecules that function together in particular ways. (Yes, folks, it's answers all the way down!)

Just as every individual living organism is the product of an unbroken chain of successful reproduction that links back to the very beginnings of life, so every living species is the result of an unbroken series of successful answers to the challenges of life. All living species are the latest in a mass of continuous, evolving streams of good answers that have flowed along time, passing through extinction filters, changing direction and evolving new forms, splitting into different streams, and sometimes merging with other good answers via hybridization.

These successful answers are the result of extraordinary amounts of evolutionary work undertaken by every species that has existed through the ages as they've struggled to survive. They're the products of uncountable rounds of trial and error, which is to say that these answers have been earned through eons of unimaginable suffering, failure, and death. Not only is it crass and stupid to throw all this work and success away by causing species extinctions, but it stands to reason that the more forms of life there are, the more possible answers to unpredictable future challenges and obstacles there will be, and thus the more chances that life will persist into the future.

We can ascribe different types of value to wild species, including financial value, cultural value, ecosystem service value, and so on. I think that one important reason that wild species have great value is not just because they are a *diverse* set of answers to the challenges of life but that many of their answers are responses to questions that *humans didn't ask or could not have asked.* Put slightly differently, nonhuman species contain answers to challenges that humans have never had to confront and may not even be able to imagine—but may have to respond to in the future.

For example, the human species and its ancestors may never have had to evolve tolerance to a toxic molecule that sometimes forms in warm ocean water, a molecule that scientists have not even discovered and named yet. But perhaps a few species of weird marine invertebrates have evolved special cells to convert this toxic molecule to a nutrient. Perhaps in the future, as ocean temperatures go up, we will need to find methods of dealing with dangerous volumes of this toxic molecule, and we'll need to turn to strange sea animals for the solution they developed. (Given the current rate of climate breakdown, this may be sooner than we think.)

For these unimaginable (to humans) answers to exist and come into being, we must honor the right of species to exist and evolve in forms that do not directly serve our immediate, definable interests which, in the greater scheme of things, are very limited. If we conserve only the species that are of obvious use to us right now, we're foreclosing on a whole lot of options for the persistence of not just our own species but of all life. Our and other species' chances of survival might soon be enhanced by an ugly, eyeless, blobby creature that lives in mud on the seafloor, a slimy fungus, or a whining, biting bug.

---

We humans have been a long time coming. The first crude single-celled life forms on Earth evolved about four billion years ago in the early ocean. It took about two billion more years for multi-cellular organisms to evolve, and yet another billion-plus years for recognizable green plants to emerge. The first backboned animals evolved five hundred million years ago, during the Cambrian explosion, and since then five major mass extinction events have occurred, the most recent being the Cretaceous-Paleogene extinction event of sixty-six million years ago, which wiped out the dinosaurs. This event opened ecological space for mammals to radiate into, but it was only tens of millions of years later that hominids, including our species, evolved into being.

*Homo sapiens*, our answer made of answers, emerged during the last bit of the period of time that geologists call the Pleistocene epoch, which began 2.58 million years ago and ended just 11,700 years ago. It was a tumultuous, difficult time for many species in many regions of the planet, particularly in the Northern Hemisphere: it's often called the Age of Ice because of the repeated episodes of large-scale glaciation that occurred throughout it.

The spread of ice across large areas was made possible about 2.8 million years ago, 300,000 years before the Pleistocene began, when North and South America were joined by the land bridge today called the Isthmus of Panama. This cut the Pacific Ocean off from the Caribbean. Warm water that had previously flowed from the Caribbean westward to the Pacific was directed northward along the east coast of North America, past southern Greenland, and on toward northern Europe; thus the massive Gulf Stream was created, which today still brings vast amounts of water from the tropics to the North Atlantic.

More water vapor evaporates from warm sea than cool sea, and warm sea thus makes more cloud and precipitation over nearby landmasses. Much of the increased precipitation that the new Gulf Stream generated over North America and Greenland fell as snow. This built up into ice sheets that reflected more solar radiation than natural vegetation previously had, which in turn further cooled the Northern Hemisphere and encouraged more yet ice sheet growth, and by about 2.7 million years ago the Northern Hemisphere was significantly glaciated. The massive ice sheets did not persist unchanged throughout the Pleistocene. Every hundred thousand years or so, the Earth tilted slightly on its axis, which changed how sunlight fell on its surface, so the ice melted and grew back through at least twenty glacial (cooling) and interglacial (warming) periods, each cycle coinciding with significant climate change and taking an average of a hundred thousand years to complete.

The last glacial maximum was reached between twenty and twenty-two thousand years ago. As much as 30 percent of the planet was covered by ice, with ice sheets being particularly large in the Northern Hemisphere: the massive Laurentide Ice Sheet covered most of modern-day Canada and a large part of the north-eastern continental United States. The Southern Hemisphere experienced less new glaciation, but significant ice sheets and glaciers still formed in Patagonia, in the highest parts of eastern and southern Africa, and in southeast Australia.

Ice-covered areas were effectively sterilized at the peak of each glaciation and then transformed back into living ecosystems as the ice melted away during interglacial times, but many parts of Earth were significantly more stable; they didn't freeze and didn't vary nearly as much in temperature and rainfall. The fact that stable areas endured as safe refuges for many species and that overall global climate change—including the advance and retreat of the ice caps—proceeded relatively slowly allowed the majority of species to survive and continue evolving throughout the Pleistocene. This included members of our genus, *Homo* (the hominids), which arose during the Pleistocene and evolved into several different species in Africa, Europe, and Asia.

Our species, *Homo sapiens*, evolved its current physical form in Africa about three hundred thousand years ago. We survived episodes of climate change by moving around the continent and learning to exploit a wide variety of food sources, although there were difficult times when our population dropped to very low levels. We developed culture, language, and simple symbolic art. Gradually we moved into Europe, where we interbred with Neanderthal people (*Homo neanderthalensis*), and then to Asia, where we may have coexisted for a time with one or two other *Homo* species before they died out.

Our species finally began colonizing Australia and the Americas about forty thousand years ago, during the last glacial period. Large volumes of seawater were held in ice, which lowered sea levels enough to expose land bridges from Asia to these continents. We wiped out some large, good-to-eat animal species in these new territories—they had evolved for millions of years in our absence, so did not fear us or know how to defend themselves against us—and changed some ecosystems through fire and agriculture, but vast numbers of wild species continued their grand work of survival and evolution despite us; the changes we wrought were relatively limited.

Throughout the evolution of our genus and species, even during the colonization of Australia and the Americas, the global number of extinctions was probably balanced out by the evolution of new species. In fact, when the last major glaciation faded away and we entered the current interglacial period that geologists call the Holocene epoch about 11,700 years ago, the Earth was probably more species-diverse than in its entire multibillion-year existence.

A lot changed for us when the Holocene began. We started becoming civilized. For the first time in our species' existence, we were able to continuously grow our populations, develop new machines, create literature, and build sophisticated systems of politics, governance, and medicine. These massive advances were made possible by one key attribute of this epoch: climate stability.

The Holocene was unusually climate-stable compared to other periods in our species' evolution, which made the seasons predictable and allowed us to develop agriculture that supported large permanent settlements—cities—in which advanced cultures emerged. Although the *Homo sapiens* of the Pleistocene were probably "intelligent" by contemporary standards and developed some of the building blocks of civilization, they didn't have the benefit of a stable climate. At the onset of the Holocene 11,700 years ago, our species was gifted an Eden filled with extraordinarily diverse, productive, nurturing ecosystems and governed by knowable, reliable seasons—and we used it.

Another huge leap forward for humanity came with the invention of the steam engine and the dawn of the Industrial Revolution around 1760. This raised living standards and enabled us to greatly boost our population, but it also sent the consumption of fossil fuels into overdrive. Humans have known about and used gas, oil, and coal for thousands of years, but until the steam engine came along we hadn't used large volumes of them. In a mere 250 years or so (less than one-thousandth of the time our species has existed!) we've gone from burning negligible volumes of fossil fuels to consuming them at almost unimaginable rates—literally millions of times faster than they formed.

Fossil fuels have become an integral part of human civilization. Look around your home: almost everything in the human-built environment—concrete, steel, timber, plastic—is made with or of fossil fuels. If the fertilizers and pesticides that we make from fossil fuels disappeared overnight, much of our agriculture would collapse. Our transport systems, medical systems, communications systems, even our arts industries, are overwhelmingly reliant on fossil fuels. They have also enabled us to convert a large percentage of the planet's natural habitat into cities, roads, farms, and plantations.

Fossil fuels have brought enormous material wealth and other benefits to humanity, but by using such massive volumes of them we've overwhelmed Earth's carbon sinks: the oceans, natural vegetation, and soils that absorb and contain energy-trapping gases from the atmosphere. We are now destroying climate stability, the fundamental element of our contemporary civilizations. We're sending into oblivion hundreds of thousands of wild species, the natural, flowing manifestations of success that have given life to the ecosystems that we've relied on since we evolved.

Simple confusion and forgetting are essential ingredients of our current stupidity. For example, we confuse money with well-being by running our economies according to theories that forget human societies are not self-sustaining: we are utterly dependent on predictable, functional ecosystems, which create the basic conditions for human life and provide material inputs and waste processing for our economies. As we damage nature, we incur costs that our financial systems usually don't recognize or account for.

Many people confuse influence with control, forgetting that having greater influence over something doesn't always mean that we have more control of it. The more we influence the atmosphere by filling it with energy-trapping pollution, the more chaotic and unpredictable the climate becomes, and the *less* control we have over the Earth system. As we degrade and destroy more natural habitat, ecosystems become less stable. This is not a problem only for wild species—it can also be a real problem for us.

I have young triplets, three girls. They're creative and lively children with many advantages, but they're growing up in a world that's obviously more depauperate than the one I grew up in. It pains me that they will not be able to have some of the valuable nature experiences that I had—some species are already gone from places I knew, and those places irreversibly changed—but in addition to that, I fear that their lives will be significantly more materially impoverished and uncertain than mine unless we rapidly arrest climate breakdown and species extinction.

My fear does not only arise from the climate research that I've read, though much of that is genuinely terrifying. It also comes from personal experiences of events that were made worse by climate breakdown, such as living through 2012's Superstorm Sandy in New York.

Sandy was barely a Category 1 hurricane at landfall, but it inflicted billions of dollars' worth of damage and made large areas of this rich, sophisticated city unlivable for weeks or months. Power networks failed, subways flooded, and buildings were destroyed. This was not unexpected. Among others, a 2008 study published in *The Bulletin of the American Meteorological Society* accurately predicted such impacts from a theoretical Category 1 hurricane that arrived at high tide, as Sandy did.

In the Rockaways area of the city a storm surge of cold seawater rushed down streets and into basements, where it shorted electrical circuits and caused heating boilers to explode, bursting pipes inside apartment block walls all the way to the top floors. Days afterward I found myself trudging up the pitch-dark, dripping stairwells of high-rise blocks as part of a grassroots volunteer effort to get food and medicine to desperate residents. Despite New York being one of the richest, best-resourced cities in the world's richest country, the authorities were overwhelmed and the Red Cross was late to the scene. Some cold, hungry, and terrified people were still locked in their apartments. They had no idea when or how they would get help as toxic black mold grew across their interior walls, making the air dangerous to breathe.

Often the mainstream media overhypes disasters, but with Sandy I felt it passed over the truly scary stories that exposed advanced societies' inability to cope with increasingly extreme weather events, even those events that are predicted and expected.

In 2018 we almost ran out of water in Cape Town, one of the best-managed cities on the African continent, because of a "once in a millennium" drought like those that are now occurring far more often than their name suggests they should. Although the drought was in line with climate model predictions for southern Africa and so could be considered "expected" under human-caused climate breakdown, it was shocking to witness how easily a relatively small, regional perturbation can lead to disaster for millions of people. I have a visceral dislike of guns—I've witnessed the damage that bullets inflict on human bodies and been shot at while making a documentary film—but at one stage the situation became so tense that my wife and I wondered if we needed one. We had ten thousand liters of water stored in large tanks in our yard; would we become the target of a thirsty mob?

The city was saved from unprecedented disaster by last-minute rain, but another "once in a millennium" drought is likely soon. Will it be next year? In a decade? Planning for your family's future feels like a very different exercise than in my parents' generation.

Hurricane Sandy and Cape Town's Day Zero drought were just little tastes of intensifying climate chaos. In recent years many millions of people around the world have had to deal with more intense storms, floods, fires, and droughts, which are causing billions of dollars of damage and plunging communities into poverty and disarray. The longer we take to act on species extinction and climate breakdown, the less we'll be able to shape our lives and those of following generations.

———

The idea that the atmosphere can retain thermal energy and so raise the temperature of Earth's surface was developed from the 1820s onward by various European scientists. In 1896 Swedish physicist Svante Arrheinus made the first serious effort at calculating how much a rise or fall in atmospheric carbon dioxide would change the planet's surface temperature. Soon after, he postulated that burning fossil fuels would cause a perceptible temperature rise, a view soon adopted by other scientists around the world.

Atmospheric science advanced rapidly in the 1900s, with estimates of the role of $CO_2$ and water vapor in determining atmospheric temperature becoming ever more refined. In 1958 the American scientist Charles David Keeling began monitoring atmospheric $CO_2$ levels at Mauna Loa, Hawaii. Despite stiff bureaucratic opposition to his work, Keeling was soon able to show that $CO_2$ levels fluctuated with the seasons—during the northern summer, the vegetation of the large Northern Hemisphere landmasses absorbed significant $CO_2$ as it grew and atmospheric levels dropped, only to rise again as the vegetation died in the northern winter—but that overall $CO_2$ levels were rising year over year in tandem with fossil fuel use.

In the 1960s and '70s other researchers, including scientists working for major fossil fuel companies, began using Keeling's data to predict future global heating. Starting in 1977, Exxon scientists developed robust mathematical models that have turned out to be extremely accurate. They predicted that human $CO_2$ emissions would cause warming of 0.2 degrees Celsius per decade—which is almost exactly what current data show—and that warming would cause discernible climate changes by the year 2000, becoming dangerous by 2050.

By the mid-1980s scientific consensus was established that fossil fuel emissions were raising the average atmospheric temperature and destabilizing the climate, but it wasn't until 1988, when NASA scientist James Hansen testified before the U.S. Congress about global warming, that the general public in the Western world became aware of the science and the potential dangers of a heating planet. At the same time public concern about other environmental matters like pollution and deforestation was intensifying; green issues were becoming politically significant in many countries, including the collapsing (and often highly polluted) Communist states.

Fearful of being outflanked by greens, mainstream politicians banded together for the first Earth Summit. Held in Rio de Janeiro in 1992, it attracted senior representatives from 178 countries and an unprecedented 117 heads of state. The summit produced several groundbreaking international agreements, including the Rio Declaration, containing twenty-seven universal principles of environmental protection, and the United Nations Framework Convention on Climate Change (UNFCC). This was the first international agreement to deal with the crisis, and has the aim of coordinating the reduction of energy-trapping gas emissions in line with scientific recommendations. The Convention has been progressively refined at roughly annual Conferences of the Parties (COPs), large meetings of the signatory nations during which progress toward the goals of the convention are discussed and new goals negotiated.

UNFCC negotiations and agreements are informed by the work of the United Nations Intergovernmental Panel on Climate

Change (IPCC), a remarkable international body composed of large numbers of climate scientists and other experts who assess the latest climate science and periodically issue reports on the effects of climate breakdown and progress made toward stopping it.

UNFCC COP meetings are often contentious, with fossil fuel interests trying to stymie agreements that would curtail demand for their products. Nonetheless progress has been made. At COP 21, held in Paris in 2015, 196 nations agreed to limit average atmospheric warming to "well below" 2 degrees Celsius above the preindustrial average, "preferably" to 1.5 degrees. These numbers represent the amount of warming that IPCC scientists think the planet's ecosystems can tolerate before they irreversibly change.

But scientists agree that we need to move much faster to combat climate breakdown, and that fossil fuel–funded misinformation and disinformation has seriously retarded progress. As documented in Naomi Oreskes and Erik M. Conway's carefully researched book *Merchants of Doubt*, fossil fuel producers, including major corporations and Middle Eastern governments, have spent hundreds of millions of dollars on sophisticated public relations and propaganda campaigns to cast doubt on the science linking their products to climate breakdown; that is, they have publicly contradicted the findings of their own researchers. (Taking climate breakdown seriously would, after all, mean that they would have to keep trillions of dollars' worth of their products in the ground.) They have financed front groups, activists and politicians around the globe, often channeling funds via circuitous channels to avoid detection. (I discovered that even South African opinion writers were receiving U.S. fossil fuel money that was being passed through a Canadian foundation and a South African think tank to obscure its origins. The opinion writers railed against climate science and promoted coal and oil burning while denigrating renewable energy as impractical, expensive, and elitist.)

Although scientific data have become ever more robust and compelling, and climate scientists ever more vocal about their findings, enough people have been confused and enough politicians paid off by Big Oil and Big Coal to ensure that large reductions in fossil fuel consumption have not yet been made.

The public might be forgiven for looking past the warnings of the scientific community. The data were mounting, and yet in most respects spring, summer, fall, and winter seemed to flow on in familiar ways. Certainly, nature faced many threats—habitat destruction, pollution, overfishing, and so on—and some countries addressed these in law, but climate change can seem abstract and distant compared to bulldozers flattening rainforests or oil slicks killing thousands of seabirds. It's only quite recently, in the last decade or so, that climate has become a major global issue again, perhaps because of the large number of record-breaking droughts, floods, fires, and hurricanes that constantly fill our media channels. Climate activists have drawn millions of young people into their movements and some countries have recently enacted climate-friendly laws and policies.

Despite these encouraging developments, dealing with the climate crisis remains complicated and difficult. Fossil fuel industry propaganda is evolving, finding new ways to greenwash their business and discourage climate action. Because fossil fuels are so embedded in our lives, it'll take a whole lot more than just swapping out a few of our home appliances for more efficient models to significantly reduce the amount we use. Despite an increase in "green" rhetoric, governments are still granting permits for new oil and gas wells and for new coal mines, which destroy carbon- and species-rich natural habitats. And banks are still pouring billions into these destructive practices; they're still, despite all the warnings from scientists, hoping to profit at our expense.

———

It's beyond the scope of this book to lay out a detailed road map for getting our societies off fossil fuels, but it's important to note that much of the technology we need already exists. Some countries have already made real progress on stabilizing and even reducing their emissions, aided by rapidly advancing renewable energy generation and storage technologies. Solar panels are more than 80 percent percent cheaper than just ten years ago, and lithium batteries for cars have gone down about 90 percent in the same period. Wind turbines are becoming ever larger and more efficient. These huge cost reductions were not anticipated during the first Earth Summit in 1992, and they have turned many skeptics into climate optimists; many researchers think we can power modern societies without wrecking the climate.

Renewable power installations are growing exponentially. Solar photovoltaic is now the cheapest source of electricity ever invented, and new investment in cleaner energy now exceeds new investment in fossil fuels. After decades of delay, the United States passed comprehensive climate legislation in the form of the Inflation Reduction Act in August 2022. The act directs $369 billion to energy and climate programs, aims to reduce US emissions by 40 percent compared to 2005 levels, and adds another $60 billion for environmental justice work. Also in 2022 Australia, the world's largest coal exporter, finally replaced their strongly antienvironmental government with a more reasonable one, and China continues to work toward becoming an "ecological society."

We also cannot ignore social and economic inequality. Rich countries, especially rich people in those countries, emit far more than average. A British study estimated that if the richest

10 percent of citizens reduced their carbon emissions to the average of the remaining 90 percent of citizens, Britain's total emissions would halve. The primary responsibility for climate breakdown lies with the rich.

Reducing emissions involves more than swapping fossil fuels for renewable energy, because about a third of emissions can be attributed to human land use, primarily agriculture and deforestation. When soil is plowed, it releases energy-trapping gases. When forests are chopped down, greenhouse gases escape from the soil and rotting vegetation. We must reduce the amount of land we use so that more land can be restored to natural habitat and become a carbon sink. Since animal agriculture uses so much land, this means cutting down on meat consumption. The growing popularity of veganism in some countries could help this, as could advances in high-tech fermentation, which will allow nutritious proteins to be produced at massive scale in vats.

Despite climate progress, scientists agree that we must do more. We have not yet turned the corner. After a temporary emissions drop in 2021 due to the Covid pandemic, in 2022 scientists recorded the highest carbon emissions ever. In June 2023 they noted a near-record increase in atmospheric $CO_2$ over the previous year; levels are now 50 percent higher than at the start of the Industrial Revolution. Fossil fuel companies continue to explore for deposits and have shown no sign of holding back on selling their vast known reserves, and some have recently retreated from earlier promises to reduce their carbon emissions.

The latest IPCC report, released in March 2023, says that the globe has already warmed by an average of 1.1 degree Celsius compared to the average temperature between 1850 and 1900. (Some credible scientists place the increase at 1.2 or 1.3 degrees C.) The IPCC calculates that $CO_2$ emissions must peak by 2025 and be cut by two thirds by 2035 if we are to stand a chance of keeping average warming below 1.5 degrees Celsius. It also states that projected emissions from existing infrastructure will take us well past 1.5 degrees Celsius; to hit the target, we must not only cease building new fossil fuel infrastructure immediately but must also decommission existing infrastructure. But there is little sign that fossil fuel interests intend to take these numbers seriously.

———

Most of the current wild species declines and ecological disruptions are not solely caused by climate change. It's seldom the only factor to blame for extirpations and ecological collapses, which are usually caused by climate breakdown acting together with other forces, such as habitat destruction, invasive species, pollution, or poaching. Climate change is usually, to use a military term, a force multiplier.

This means that we can meaningfully reduce the impact of climate breakdown by curtailing the other causes of harm. We can reduce the likelihood of extinctions and increase ecosystem resilience by restoring natural habitats, combating invasive species, cutting pollution, and stopping the overexploitation of wild species. We can also make data-informed predictions about how and where wild species are likely to move as the globe warms, and create corridors of natural habitat leading to those places.

Habitat restoration, now often called rewilding, can be useful almost everywhere and at any scale. For example, replacing your lawn with locally native wild plants can support those plants as well as a rich variety of birds, insects, and smaller animals. This is no less important than protecting huge areas of wilderness in national parks; lawn grass is the United States' largest irrigated "crop," covering more land than irrigated corn, wheat, and fruit orchards combined. Turning even a small percentage of lawns into wildlife habitat can have a measurable positive effect on insect numbers—as well as cutting down on water, pesticide, fertilizer, and fossil fuel use—and a few gardeners and landscapers are beginning to do just that.

At the other end of the scale are efforts to rewild and conserve millions of contiguous acres of land to sequester carbon and conserve close-to-complete species communities, like American Prairie, a prairie-focused preserve in central Montana. The preserve aims to conserve over three million contiguous acres of public and private land, allowing bison and other native species to migrate as they would have before the area was cut up by cattle fences. Good progress is being made in linking national parks with large cross-border corridors in many regions of Africa, thus creating massive conservation areas which allow wild species to migrate and ecosystem restoration efforts to expand. Many parts of Europe are returning to wilderness as agriculture's footprint shrinks on that continent.

Marine conservation has also made huge advances in recent years. Many countries have acknowledged the usefulness of marine protected areas (MPAs), where fishing and other potentially destructive activities are limited or entirely forbidden, and have proclaimed enormous tracts of their territorial waters as such. The Papahānaumokuākea Marine National Monument grants full protection to over 1.5 million square kilometers of U.S. territorial waters west of Hawaii. The Ross Sea MPA, off the Antarctic coast, covers an area larger than Mexico.

In December 2022, at the fifteenth UN Biological Diversity Conference (COP 15) in Montreal, most of the world's countries agreed the so-called 30x30 deal. Although the agreement was nonbinding, its signatories agreed to conserve 30 percent of their land and seas, restore 30 percent more, and deliver $30 billion in funding to poorer countries to help them conserve natural habitat. The details of what constitutes conservation and restoration still need to be hammered out, and there are concerns about how it could erode Indigenous peoples' land rights, but it marks a potentially massive step forward in conserving nature.

It's imperative that climate breakdown be prevented for traditional habitat conservation and restoration to succeed, because high levels of warming will likely scramble and wreck even the most carefully managed natural areas.

But it's important to recognize that many so-called climate solutions can have their own damaging impacts on nature, usually by directly destroying large areas of habitat. Solar power companies are already bulldozing hundreds of thousands of acres of wildlands in the American Southwest for panel arrays, threatening many species; it's initially cheaper than placing panels on individual rooftops, and allows a few influential players to control and profit from the industry.

Massive tree-planting projects, ostensibly to sequester carbon and to restore nature, are replacing natural areas around the world with monocultures of exotic trees, thus driving out wild species and consuming huge volumes of water. Huge tree-planting projects are often paid for by emission offset schemes, which offer people or companies the opportunity to keep emitting dangerous gases in return for a payment that is supposed to fund carbon sequestration. (Most of these have recently been found to be fraudulent.) Politicians are attracted to these projects for the photo ops and fawning media coverage they generate—who doesn't love having their picture taken with a cute little tree?—even though the projects often fail in the long run and drive poor rural people from their land.

Biofuels made from plants are touted as an eco-friendly alternative to fossil fuels, but they can cause huge environmental damage. A stunning 40-plus percent of the U.S. corn crop is used to produce ethanol to power cars and trucks. This uses tens of millions of acres of land and extraordinary volumes of agrochemicals and fossil fuels, and because the process of converting corn to liquid fuel is highly inefficient, it results in almost no climate benefit.

These sorts of land-hungry programs are already driving a wedge between species conservationists and climate activists, who should be working together to ensure that we don't kill the planet to save it. Some activists and energy policy wonks have unfortunately not yet got the message that climate "solutions" that involve destroying large areas of natural habitat are not solutions at all.

———

Meaningful progress on climate goals is being stalled in many countries by politicians who are funded by fossil fuel interests and polluting industries. Necessary action won't happen unless they are pushed to act by their constituents or replaced, which requires voting for the right candidates and effective activism. Banks and investment firms are still putting billions of dollars into exploring for new fossil fuel reserves and boosting their production and consumption. Governments still subsidize fossil fuels with billions of taxpayer dollars. This money needs to be redirected to reducing emissions and protecting natural habitats.

Individuals changing some of their consumer choices, like replacing their gas-powered lawn mower with an electric one, will not achieve the necessary change at the speed and scale required. Reusable drinking straws will not save us from climate Armageddon. Millions of people must work together to change large systems of production and consumption, of settlement and transportation, and do it fast.

Activism comes in many forms, and many of the most effective climate activist groups use nonviolent civil disobedience alongside conventional political lobbying and community organizing to drive their agenda into the mainstream. In the United States the Sunrise Movement has been instrumental in shaping pro-climate legislation, working within and outside the halls of Congress. Other groups like 350.org and Third Act have organized large public demonstrations and effective disinvestment campaigns. Numerous branches of the Extinction Rebellion movement have blocked traffic and conducted large demonstrations in the United Kingdom and many European countries, putting climate and species extinction center stage in the media.

Scientists and academics are increasingly speaking out in public and joining demonstrations, rejecting earlier notions that "good" scientists should stay out of politics and avoid being seen to take sides on important issues. Leading climate modelers and medical researchers in Europe and North America are being arrested in significant numbers as they take part in civil disobedience actions under the umbrella of Scientist Rebellion.

And lawyers are joining the climate movement, too, both by acting on behalf of activists and also by advancing the new legal concept of nature's rights, which is gaining traction in many countries. Nature's rights posits that wild species and natural ecosystems and places—like rivers, forests, or mountaintops—have intrinsic rights and should be regarded as legal persons, capable of legally defending themselves against human violation. This represents a profound paradigm shift in law and society, and in some countries natural entities have already won cases against their human persecutors. Rivers in countries as diverse as India, New Zealand, Canada, and Colombia have been granted legal personhood with certain rights, like the right to exist, to flow, to live free from contamination, to be protected and restored.

It's not enough that climate science is clear and that we have the tools to reduce dangerous emissions. It's not enough that millions of people around the world are rightfully concerned about climate breakdown and want to prevent it. Resolving the climate crisis requires intelligent, courageous people to gain and use political and economic power.

———

The early stages of researching and writing this book inspired me. Climate breakdown and the natural world is a subject that could engage me for years, to which I could apply knowledge and insight from my various careers and passions. It seemed like a project perfectly made for me.

My inspired state didn't last long. Although I've long known the basics of climate change and its links to the extinction crisis, I was not prepared for the heap of terrifying facts that I unearthed as I plowed into piles of scientific publications and interviewed experts from around the world. The stories of disruption and irreversible change became overwhelming, so numerous they would require many lifetimes to take in. I realized that my book couldn't be grandly authoritative or all-encompassing; I'd have to be humble, accept uncertainty, and write about just a few species and places in the hope that they'd adequately illustrate the sorts of things that are happening everywhere.

I concentrated on places I knew or had lived in before, but I also wrote about some new places I visited for this project, like the deserts of the American Southwest, the forests of Puerto Rico, and the wildlands of southeastern Australia. These trips were fascinating and worthwhile, but they weren't easy. I saw ecosystems thinning out, collapsing and morphing before my eyes, species fading and stuttering into extinction. I stood with experienced ecologists as they ran out of words to describe the effects of climate breakdown. Tough men fell into tears as they showed me what was happening to the creatures and living systems they'd spent their lives studying and conserving.

I saw many things I'll never forget, but a few encounters fundamentally changed me, like one beautiful blue-sky-and-white-cloud day in late February 2020 when I drove up to Australia's famed Kozsciousko National Park in the Snowy Mountains, halfway between Sydney and Melbourne. Much of it had burned six weeks earlier, but in the interim there had been good rain, and I wanted to see how it was regrowing after the intense fires that had made world news for weeks on end. Many of Kozsciusko's tree species are fire-adapted and should bounce back rapidly after a burn.

A couple of miles into the park I pulled my rental car over to the side of the Snowy Mountains Highway to get a closer look at a large swath of burned forest. I'm no stranger to wildland fire, and it doesn't scare me—I grew up in and live in habitats that naturally and regularly burn—but I was stunned by what I saw. A dense mass of thousands, millions, of dead trees filled the landscape from horizon to horizon. The forest was the brown of dead wood and the black of charcoal. It smelled of ash and burned oil. I had never seen a place so thoroughly and extensively burned.

I wandered a few hundred yards off the road, into the dead trees, and was surrounded by total silence. No birds sang. No insects buzzed. There were none of the fire refugia, the small unburned patches that you normally find in a post-fire landscape, where many species survive blazes. I found a few small rock outcrops; they were littered with half-inch thick flakes of rock, some larger than my face, that had exploded off the boulders from the heat of the fire. Almost nothing was regrowing beneath the trees because the fire had been so hot for so long that it had vaporized the organic matter in the soil, including, apparently, the seedbank. There was just dry, gray-brown sand on the forest floor.

There were some greenish arteries running through this ruined land. Threadbare carpets of grass were emerging along the small creeks, now choked with dark ash and sediment. Grass also sprouted along the road verges, immediately next to the asphalt—its seeds had been just far enough away from the burning trees not to be killed—but these strips of green only brought more death: I have absolutely no idea how, but a few marsupials had survived the fire and, obviously starving, had made their way to the new roadside vegetation and been run down by cars. The stiff, hairy carcass of an Eastern Gray Kangaroo lay bloating on the asphalt near me, unbothered by scavengers.

I felt like I was witnessing the future in the incinerated forest of Kozsciusko, like I had passed into a place so new I hadn't fully caught up to myself yet. It felt unarguably real yet profoundly unfair, like the feeling I've had when looking into the faces of dead people who were not ready to die, those who fought death until the last second. It was grand, epic, complete in its horizon-to-horizon endlessness and thoroughness, but not fully comprehensible, and in no way beautiful. The only words I could find in response to the scene were: *We really can fuck this all up. All of it.*

I left Australia and flew back to South Africa as the news of a new virus, SARS-CoV-2, was beginning to spread around the world. The Chinese people on my flight were all wearing masks, which made the rest of us uncomfortable. Shortly after I arrived home, the South African government placed the country into extreme lockdown. My wife developed long Covid, which confined her to bed for months and upended our family's life for well over a year. Although previously healthy, she developed scary heart problems, skin problems, tinnitus, and digestive disturbances, many of which the doctors could not explain or even admit to not understanding. The unpredictability of the virus and disease and the often-clueless response, including the widespread denial of its dangers, reminded me a lot of the current situation with climate breakdown.

As a child I learned about Medusa, a powerful winged monster from ancient Greek mythology with a human female body and living snakes for hair. Medusa's visage was horrifying—she is often depicted with deathly, staring eyes and huge, sharp teeth—and any regular human who looked at her face would instantly be turned to stone. I'm no ancient history buff, but her story comes to me often as I write. I feel like a hapless Greek soldier who has been commanded to confront her again and again, and then again. I see unnerving signs of serious change everywhere now, in the sudden arrival in my neighborhood of new bird species from distant places, in the jarring yet easy warmth of unusually hot days in winter, in the new science showing that climate shifts are occurring more rapidly than expected, in the media reports about extreme storms, fires, and droughts around the world.

I often find myself psychologically turning to stone, unable to fully grasp what I'm reading or seeing, unable to think clearly or write coherently, even for weeks at a time. Sometimes I can intellectually understand something about climate breakdown, but it feels so abstract as to be meaningless, even if it is right in front of me. I am not alone in this. Many people who study climate breakdown report similar responses and more; anxiety, depression, chronic fatigue, and other awful afflictions. "Climate dread" is a real and growing phenomenon.

I now better understand those climate researchers who feel separated from society because they cannot comprehend why so many people don't see their findings as heralding an emergency. It's profoundly alienating to carry and communicate important knowledge that the people around you won't act on—and that even you struggle to act on, because you must make a living in an economy whose regular operation causes the problem that you've identified. To survive, you must make things worse.

Psychologists tell me that the less agency you have during a traumatic experience (the less power you hold to affect the outcome), the more severe and long-lasting the trauma can be. Individual scientists, researchers, and writers can feel almost completely powerless when confronted with the enormity of climate breakdown and the ongoing failures by those in charge to substantively deal with it.

Psychologists also tell me that people sometimes respond to a traumatic experience by dissociating during the experience, or splitting it up in memory. You might vividly remember one aspect of a car accident, for example—what it looked like, smelled like, or sounded like, but not all those things together. You might easily recall the beginning of an incident but not the end, or the emotions you felt during an incident but no visual details of it. Psychologists can work with clients to gradually reintegrate the various aspects of a traumatic experience, remember and bring them together into one story, so they can more fully understand it, process it, and heal.

But climate breakdown may be too big, complicated, and powerful to bring together coherently. Although I've researched and written this book about it, I still haven't figured out what to tell my children about it, and am not sure how to psychologically align my understanding of it with regular things like getting them to school on time, buying groceries, and planning vacations. How do you keep your mind on day-to-day tasks while being aware of unfolding worldwide chaos and destruction?

In the old Greek stories Medusa was ultimately killed by the hero Perseus, who used a mirrored shield to see her without having to look directly at her face, and then cut off her head with his sword. There's something to be gleaned from that—perhaps regular humans simply aren't able to look too directly or completely at the results of our gas emissions without suffering debilitating psychological disturbances. Maybe it's more practical to understand climate breakdown through smaller stories, like the ones in this book, than by trying to wrap your mental arms around the whole apocalyptic disaster.

Although I haven't neatly solved the problem of how to think about climate, I have found that focusing on it has made it easy to liberate my attention from things like professional sports and social media scandals. Who won the big game or said something "woke" or "problematic" online? I care less and less all the time, and it's wonderful.

I'm also increasingly sure that if we persist in structuring our economies according to mainstream models of economic "development" and "growth," many established ways of doing things will rapidly (and often unexpectedly) become impossible, either because a critical mass of people decides that they're morally indefensible or because the climate and social-ecological systems that undergird them fall apart.

It's clear that we have a choice: we can stay on the business-as-usual train, suffer the predictable as well as the unpredictable consequences, and inflict more irreversible damage on the natural world, or we can make meaningful changes to our lives and societies to curtail our heat-trapping gas emissions, the extinction of wild species, and the dismemberment of ecosystems.

Taking the latter route seems obviously right, although it's far easier said than done within social and economic systems that have been set up to efficiently trash the planet, systems that we're so embedded in that it's hard to see where they begin and end, or where best to start changing them.

Sometimes it seems hopeless. We've already done so much damage. Solving the climate crisis is an almost incomprehensibly massive job; we must change so much so fast, and there are no silver bullets, no single actions, policies, or technologies that'll help us achieve our goals. We must retain unity of purpose while being bombarded by a cacophony of conflicting opinions about what's needed, many of them cunningly crafted by bad actors to throw us off track. We must take on and beat extremely strong, entrenched fossil fuel profiteers who will do almost anything—including murder their opponents, foment revolutions and wars, and destroy their own families' opportunities to live on a stable, thriving planet—to retain their power.

And there is a real possibility that we will fail.

All this is true, so it's no wonder that many people succumb to climate "doomism," slumping into indifference and inaction because they believe that global disaster is inevitable.

I've sometimes fallen into doomism myself, but I never remain in this state for too long because of other things I know to be true. For example, conservation biologists have learned a lot about saving species and repairing ecosystems, and have successfully restored many endangered species and trashed habitats to health. Endangered species are not destined to vanish, and ecosystems can resurrect themselves, sometimes rapidly, if the correct conditions are created.

We've developed affordable, reliable technologies that allow us to radically reduce the amount of fossil fuel we burn and the energy we require to maintain a good standard of living—everything from solar panels to advanced batteries and efficient electric stoves—and these are getting cheaper and better all the time.

New technologies can become ubiquitous in the blink of an eye, just as the internet and cell phones moved from being fringe technologies used by a small minority of people to everyday essentials in just a few years of my life. Vicious cycles can make things very much worse very quickly, but virtuous cycles can make them better almost overnight, too.

Dominant moral values and habits can shift over the span of a single generation, and we can swiftly stop doing harmful things to ourselves when necessary. I remember clearly when cigarette smoking was normal almost everywhere, most adults seemed to smoke, and conventional wisdom held that it was almost impossible to stop because nicotine is so addictive. Now cigarettes are almost invisible in the public places of many countries, and I often go to large social events where no one smokes. Vegetarianism was rare in the communities I lived in when I was younger. Now vegan restaurants exist and every supermarket has a wide variety of plant-based foods on sale.

Among the strongest forces preventing countries from rapidly reducing their energy-trapping gas emissions are the entrenched fossil fuel interests mentioned above. But governments and institutions that appear to be unassailably strong can lose power and collapse unexpectedly quickly under pressure—just ask anyone who lived through the end of formal apartheid in South Africa or the disintegration of the Soviet Union. Fossil fuel corporations and corrupt petrostate governments are not invincible.

None of these changes happened easily; they happened because people with a vision of a better way of doing things worked, sacrificed, and collaborated to achieve their aims.

We must change a lot about the way that we make things and use things, how we consume and dispose of things, to prevent the worst effects of climate breakdown. This is a colossal, intimidating job, but it means that opportunities for making a positive difference are abundant everywhere. There's something useful for everyone to do, no matter where or who they are.

It's not inevitable that we will one day live in a stable, bountiful biosphere even if all of humanity agrees that it would be a nice idea and acknowledges our ecological sins. The Earth system doesn't care about our thoughts and feelings, or if our intentions are pure. Our actions matter, but even if those of us who care work hard and smart, there is no guarantee of success; we might not prevent mass extinction and large-scale ecological breakdown. But what better goal is there to strive toward than the continued flourishing of life itself, in all its miraculous forms?

# ACKNOWLEDGMENTS

A book like this doesn't get written without support and input from many, many people, and in my case that support and input began decades ago, when I was a youngster. A few people particularly encouraged my budding nature-nuttery, including Tony Harris and Wulf Haacke of the (then) Transvaal Museum; Tony spent days showing my friends and me how to identify, catch, and ring birds for research, and Wulf taught us reptiles. The Neser family brought me along on numerous hikes into the wild and found it completely normal that a kid would want to know the scientific names of creatures and bring them home alive in bags and jars because, well, they did too. Christine Lambrechts supplied transport, chocolate, and off-color humor for many memorable birding expeditions to sewage plants, wetlands, and far-flung corners of the South African bush.

I'll forever be grateful to the good people of the Middlebury Fellowships in Environmental Journalism, who granted me a fellowship in 2007 that helped kick-start my international career in wildlife and environmental writing. I met Chris Shaw, Sue Kavanagh, and Bill McKibben at Middlebury and they've been valued friends and supporters ever since. I remain inspired by fellow grantees, including Sasha Chavkin, Phil McKenna, Carolyn Kormann, Heather Smith, Andrew Mandimbonyani, and Forrest Wilder: it's good to know that I'm not really alone in this often-lonely work! Roger Cohn of Yale Environment 360 has given me valuable opportunities and I've learned a lot while producing articles for that publication.

*The End of Eden* would not have been possible without the vital contributions of numerous naturalists and scientists—mostly biologists of various subspecies—who shared data, publications, valuable perspectives, and time in the field so I could learn about the organisms, places, and ecosystems they've spent their lives researching. (Their work really matters!) Most are named in the Notes on Sources section, but I'd additionally like to thank Glenn Walsberg, Jessica Tout, Eric Vanderduys, and Jamie Pittock for their contributions, not all of which made it into the final text.

I'd like to thank Jesse Breytenbach for her beautiful line drawings, and the following people who helped me source photos and/or generously provided them for use in the book: Dan Bergeron, Lara Anderson, Jeff Garnas, Caroline Kanaskie, Alexandra Bukvareva, Sergei Khomenko, Igor Shpilenok, Jacob Drucker, Rachel Kingsley, Zach Pezzillo, Paul Baker, Kristi Fazioli, Sherah McDaniel, John Magera, Tom White, Jan van de Kam, Mark S. Graham, Timm Hoffman, Dave Ward, Conor Eastment, Stephanie Tate, Mia Hoogenboom, Ed Gullekson, Ruben Jenkins-Bate, and Roman Dial.

Many friends, both old and new, accommodated me and/or took me birding while I was on reporting trips. Among them are Anna Elbers and Lornie Phillips, Jonathan Franzen and Kathy Chetkovich, Ceal Klingler, Jacob Forman, the Gilfillan-Daniels clan, Gordana Pozvek, Marian Neal, and Michelle and Frans Lombard.

My mother, Betty Welz, did a sterling job of tracking down typos in drafts and helped me clarify my writing. Sarah Geline, Carey McKenzie, Chris Shaw, and Rupert Koopman read drafts and made useful comments on them.

Finally, I extend heartfelt thanks to my editor, Anton Mueller, for his ideas, comments, and especially his enduring support as I grappled with this book through the Covid pandemic, and my agent, Aoife Lennon-Ritchie, for her sustained and tactful advocacy.

I've doubtless left some names out of this section. Apologies if you helped and don't see yours here! Apologies also if you provided a fact or insight that I have misunderstood or inadvertently misrepresented. Although I've made considerable effort to ensure the accuracy and truthfulness of the contents of *The End of Eden*, it's in the nature of making books that a few errors will have slipped in. I take full responsibility for them and will post corrections on the book's website as I become aware of them.

# A NOTE ON SOURCES

The facts, explanations, and scenes in this book are drawn from a range of sources. These include my own decades of "naturalisting" around the world, my education in zoology, botany, and ecological economics at Rhodes University and the University of Cape Town (both in South Africa), numerous email exchanges, interviews, and field visits with naturalists and scientists, and hundreds of peer-reviewed scientific papers, books, official and academic reports, and credible websites.

Some scenes are based on information from more than one type of source, but in all cases—including the composite scenes that I encourage the reader to imagine and the scenes that I have imagined myself—I have based my writing on credible facts as much as possible. I didn't witness Hurricane Maria killing Puerto Rican Parrots, for example, but I wrote a scene describing it based on my posthurricane visit to El Yunque National Forest, first-person accounts of Puerto Ricans who went through the storm, and official scientific reports on it. The ghost of the Cheetah in chapter 6 is most definitely imaginary, but the ecosystem she walks through exists, and my description of it is based on my personal experience of similar landscapes in southern Africa and Namibian experts' descriptions and photographs of the precise area.

A short, chapter-by-chapter outline of sources follows. Please see this book's website, www.theendofeden.com, for an extensive list of detailed notes, references, and errata.

INTRODUCTION

Many details about New York parks are based on my years of living in the city between 2008 and 2014, during which I birded extensively and came to know Brooklyn's Prospect Park especially well. Information about diseases and new insect pests was obtained from various government websites, scientific publications, and conversations with American entomologists and horticulturists. Todd Forrest of the New York Botanical Garden was particularly helpful in describing recent climate-related changes in the area. General facts about climate history and increased $CO_2$ levels were obtained from mainstream scientific sources including National Aeronautics and Space Administration (NASA) websites and Intergovernmental Panel on Climate Change (IPCC) reports. The mainstream scientific view is that global biodiversity has generally risen over the past hundred million years, peaking in the last few thousand, just before settled human civilization began to spread, although a minority of scientists think that it may have been at roughly that level for a long period before then. There is no evidence that overall biodiversity was in decline at the dawn of human civilization.

CHAPTER ONE: ENERGY, WATER & TIME

The description of the Mojave Desert and its plants and animals is based on my book-research trip to California in September 2019, during which I spent a few days walking in Death Valley National Park and Joshua Tree National Park. Details of Joseph Grinnell's life and work were gleaned from various sources, including the websites of the Museum of Vertebrate Zoology (University of California, Berkeley) and the Grinnell Resurvey Project, based at the aforementioned museum. Information about the current decline in Mojave bird species and the status of small mammals in the region was obtained from peer-reviewed scientific publications authored by participants in the Grinnell Resurvey Project and in-person interviews conducted at UC Berkeley with Steven Beissinger and Eric Riddell. The late Barry Sinervo walked me around his Side-blotched Lizard research site near Los Baños, California, explained how increased temperatures are killing off some California lizard populations, and intro-duced me to his UC Santa Cruz students' lab work with salamanders. Facts about the Joshua Tree were taken from various peer-reviewed scientific publications. Descriptions of the Kalahari region and the Southern Yellow-billed Hornbill come from my own experience. Andrew Clarke's book

*Principles of Thermal Ecology* was especially useful as a basis for understanding and writing about living organisms' relationships with energy and water; I benefited from many other sources, too, including E. C. Pielou's *The Energy of Nature* and the International Union of Physiological Sciences' *Glossary of Terms for Thermal Physiology* (3rd edition). Facts about Southern Yellow-billed Hornbill breeding, population collapse, and reaction to increased temperatures were based on published, peer-reviewed research first authored by Nicholas Pattinson at the University of Cape Town's Percy FitzPatrick Institute of African Ornithology, as well as discussions with the very helpful Susie Cunningham of the same institute.

CHAPTER TWO: PLAGUES & DISEASES

The description of the dying Moose is based on a video clip taken by Jacob DeBow. Pete Pekins, professor emeritus at the University of New Hampshire, generously provided a wealth of context, facts, and scientific papers about Moose, forest ecosystems, and Winter Ticks in New England during online video interviews and by email. I am generally familiar with forested ecosystems in the northeastern United States, having traveled to various parts of the region while living in New York City. Facts about ice came from various credible sources, including *Principles of Thermal Ecology*. I was first taken into a patch of Pine Barrens habitat in 2008 by the late David Burg, an inspirational New York naturalist and dear friend. Facts about that ecosystem were taken from authoritative U.S. government and nonprofit conservation websites. Jeff Garnas, Kevin Dodds, and Matthew Ayers shared enormously useful information (including many scientific papers) about the Southern Pine Beetle and other tree-consuming insects and diseases via online video calls and email. The account of the 2015 Saiga antelope die-off and its causes was based on contemporary news accounts, reports by field conservationists, and published scientific articles, including "Saigas on the Brink: Multidisciplinary Analysis of the Factors Influencing Mass Mortality Events," a multiauthor paper published in *Science Advances* in 2018. The Hawaiian Honeycreeper section is based on peer-reviewed scientific papers and information provided by Hawaiian scientists, especially Melissa Price and Christa Seidl of the University of Hawaii. Information about diseases contained in the Arctic permafrost comes from published scientific papers and news reports from well-regarded mainstream outlets.

CHAPTER THREE: EXTREME WEATHER

I traveled to Puerto Rico in late September 2019 and stayed some days at the El Verde Field Station in El Yunque National Forest. During that time I was introduced to the forest's ecology and its reaction to hurricanes by Jess Zimmerman of the University of Puerto Rico, who leads the Luquillo Long-Term Ecological Research Project in the forest. Tana Wood of the U.S. Department of Agriculture Forest Service took me to her research site, where she explores the effects of raised temperatures on the forest, and explained her work. Tom White of the U.S. Fish and Wildlife Service, who heads the Puerto Rican Recovery Program, showed me around the parrot captive breeding facility, described his experience of Hurricane Maria and its aftermath, and gave me large amounts of information about the birds from his many years of working to save them from extinction. All three of the aforementioned people provided me with useful scientific papers. Additional facts about the parrots, Caribbean geology, Puerto Rico, human presence on the island, deforestation, and other species came from peer-reviewed scientific papers and official reports. Facts about hurricanes and the physics of water, and details about Hurricane Maria's track, intensity, and effects came from *Principles of Thermal Ecology* and official U.S. government reports, including that of the National Hurricane Center. Vanessa Mintzer of the Galveston Bay Dolphin Research Program told me about their research on the Bottlenose Dolphins of the bay and the effects of Hurricane Harvey in an online video meeting and emails. Pádraig Duignan of the Marine Mammal Center conveyed additional information about dolphin physiology and Freshwater Skin Disease in an online video call and emails. I found facts about dolphin evolution in peer-reviewed scientific papers. John Magera, manager of the Attwater Prairie

Chicken National Wildlife Refuge, generously helped with information about the bird and Hurricane Harvey's effects on its population via a video call and emails. Further facts about Attwater's Prairie Chicken and its habitat was obtained from scientific papers, official U.S. government reports, credible news articles, and publications by conservation nonprofits, including the Nature Conservancy. Also in September 2019 I visited the Florida Keys, staying a few days on Big Pine Key to see Key Deer in their habitat. The deeply knowledgeable Chris Bergh of the Nature Conservancy guided me around the island, showed me the effects of recent hurricanes on its ecosystems, and pointed me to much useful information about the region and its species.

CHAPTER FOUR: MORPHING MIGRATIONS

I base my description of the Banc d'Arguin National Park on satellite images viewed via Google Maps and photographs taken by conservationists working there. Jan van Gils of the Royal Netherlands Institute for Sea Research provided a wealth of information about the Red Knot, which he's spent a career studying, via online video call and email; more facts about that species were gleaned from peer-reviewed papers and other published research by him and his collaborators. Eva Kok of the same institute provided additional information via video call and email, including details of Red Knot behavior in the Netherlands. Facts about Earth's magnetic field and bird navigation methods come from a wide range of peer-reviewed scientific papers and books. My description of White Storks is based on my numerous encounters with the species in South Africa. Details of the German countryside and breeding site were taken from Google Maps satellite images and photographs and written reports by local scientists. Facts about White Stork migration were taken from numerous credible published sources, including those by scientists working with Sociedad Española de Ornitología (SEO Birdlife). Details of the Blackcap Warbler, Richard's Pipit, Willow Warbler and other songbird migrations were found in peer-reviewed scientific papers, many of which were recommended to me by Magda

Remisiewicz of the University of Gdańsk's Bird Migration Research Station. Madhusudan Katti of the College of Natural Resources at North Carolina State University pointed me to his calculations regarding *Phylloscopus* warblers and the rate at which they consume insects.

CHAPTER FIVE: FIRE

In 2020 I drove over a thousand miles across southeastern Australia, from Brisbane to Melbourne, and then flew to Kangaroo Island near Adelaide. The trip took three weeks and allowed me to see many natural habitats, as well as areas that had recently been burned in the much-publicized Black Summer bushfires. Many Australian scientists and conservationists were generous with their time and expertise, often accompanying me to sites of interest. Luke Pearce, then a fisheries manager with the New South Wales Department of Primary Industries, drove me to Mannus Creek in the Bogandyera Nature Reserve, guided me around the area, described the deluge of ash that he witnessed moving down the creek, and provided useful information about the Macquarie Perch and its conservation. I found additional Macquarie Perch facts on Australian government websites and in published scientific papers. The description of the upper reaches of Mannus Creek was derived from Google Maps images. I got details of the Green Valley–Talmalmo fire from credible news websites, Australian government reports, and information emailed to me by Reynir Potter of the New South Wales Rural Fire Service. I learned more than I had known before about the physics of fire from *Principles of Thermal Ecology* and various credible websites. The paragraphs dealing with the evolution of green plants, the origin of fire, and the evolution of Australian vegetation are informed by a diverse array of scientific papers, Tim Low's book *Where Song Began*, and conversations and emails with Australian ecologist Robert Kooyman. (Kooyman attempted to take me to see Nightcap Oak trees in the wild, but we were stymied by downpours and fallen trees across the road.) Mark Graham, then an ecologist with New South Wales's Nature Conservation Council, shared his knowledge, introduced me to some extraordinary tracts of Gondwanan forest (burned and unburned), and showed me some of its amazing animals and plants, including the Superb Lyrebird. He's kept me updated on the fate of several key species since my visit. Another fine Australian, Philip Zylstra, shared some of his vast knowledge of eucalypts and fire with me. Much of my understanding of fire has been influenced by personal experience of fires and their aftermaths in various African habitats and my education at Rhodes University and the University of Cape Town. Paragraphs about El Niño and the Indian Ocean Dipole were based on credible websites and published scientific papers. I learned the basics of the Ponderosa Pine from Sylvester Allred's book *Ponderosa: Big Pine of the Southwest*, and more detailed information about the tree from various peer-reviewed papers. I learned a lot about the history of Californian forest management and the impact of fires from Hugh Safford (then regional ecologist with the USDA Forest Service), who showed me around the South Lake Tahoe area in September 2019. I spent a day with Art Shapiro of UC Davis driving a long route through the Sierra Nevada to see different habitat types and talk about the impacts of climate change on Californian butterflies, which he has studied for decades.

CHAPTER SIX: FERTILE AIR

Laurie Marker and Bruce Brewer of the Cheetah Conservation Fund helped me with detailed information about Namibian Cheetahs via a video call and emails. Ross Barnett, paleontologist and author of *The Missing Lynx*, told me a lot about Cheetah ancestors and pointed me to useful scientific research on Cheetah evolution. Paragraphs about early ecologists, including Humboldt and Clements, are based on numerous websites, books, and scientific papers, as is the section on the origins of photosynthesis and the different forms of photosynthesis. My knowledge of southern African savanna ecosystems and carbon fertilization comes from personal experience, my university education, interactions with southern African ecologists, and many scientific papers—the work of William Bond of the University of Cape Town and his collaborators being particularly informative. The description of the habitat at Otjozondupa is based on descriptions by Namibian ecologists and my own experience with the same plant and bird species elsewhere in southern Africa.

CHAPTER SEVEN: SEA CHANGE

Facts about the Great Barrier Reef are from Australian government and United Nations websites and Google Maps. The account of the prehistory of the ocean, its role in absorbing carbon, and ocean acidification come from various peer-reviewed scientific papers. I live near an African Penguin colony and have learned much about the birds via personal observation, published science, and local ornithologists. Christina Hagen, who works on African Penguin conservation with BirdLife South Africa, has been especially helpful in recent years—I traveled with her to the new "artificial island" colony, which she has spearheaded. Information about Green Turtles on the Great Barrier Reef comes from various scientific papers and credible websites. I have personal experience of kelp forests in South Africa and on the west coast of North America, but further information about that habitat comes from scientific publications. The account of the Sunflower Sea Star and its habitat comes from conservation websites and scientific publications. Facts about Sea Star Wasting Syndrome come from numerous peer-reviewed scientific papers. Details about coral bleaching in Australia come largely from sources (including scientific papers) suggested by Terry Hughes of James Cook University in Queensland. Maoz Fine of the Hebrew University of Jerusalem generously explained his research on heat-resistant corals in the Gulf of Aqaba to me via video call and helped me with some coral facts; I have also used his published research.

CHAPTER EIGHT: STABLE / UNSTABLE

I live on the Cape Peninsula, and my description of the peninsula, fynbos vegetation, fynbos fires, and Cape species is largely based on my own experience. Darwin's words about this area come from his *Journal of a Voyage around the World*. Information about Darwin's finches comes from scientific papers. Facts about the Cape Floral Region come from various published sources, including those by the Botanical Society of South Africa and Conservation International. The research of many ecologists has generated the current understanding of how the fynbos came to be so diverse—it's impossible to name them all here. Jasper Slingsby of the University of Cape Town is researching fynbos plant responses to climate change, and much of my understanding of this comes from his published work and numerous conversations with him. Paragraphs about general global warming are based on sections of the IPCC reports and conversations with NASA climate researcher Peter Kalmus. The section on the Equatorial Undercurrent and its moderating influence on the Galapagos is largely based on a 2022 paper lead-authored by Kris Karnauskas of the University of Colorado, Boulder. The Alaska section of the chapter is obviously based on my own travels there, and also on the scientific research of Roman Dial, an ecologist at Alaska Pacific University.

CHAPTER NINE: CONCLUSION

Much of this chapter is based on my own experience. The account of the history of climate science comes from various U.S. government websites (including those of NASA), IPCC reports, and Naomi Oreskes and Erik M. Conway's book *Merchants of Doubt*. Other facts are largely sourced from peer-reviewed scientific papers and official websites, and are detailed in the reference section on this book's website.

# INDEX

# Index

# Index

# Index

# A NOTE ON THE AUTHOR

ADAM WELZ is a widely traveled environmental writer, photographer, filmmaker, and nature conservation consultant. His work has appeared in the *Guardian*, Yale Environment 360, the *Atlantic*, Ensia, and many other outlets worldwide. He's a recipient of a Middlebury Fellowship in Environmental Journalism and currently lives in Cape Town with his wife, Sarah, and their triplet daughters. You can find him at www.adamwelz.net, on Twitter and iNaturalist @adamwelz, and on Instagram @adamwelz.wild.